Leaning Over To Hear

Leaning Over To Hear

41 PIECES ABOUT THE CHILD IN EACH OF US

Ann Giacalone D'Ippolito

The Point Press
Cape May Point, New Jersey
2008

pTp
The Point
Press

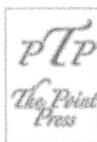

LEANING OVER TO HEAR
41 Pieces about the Child in Each of Us

ISBN 978-0-578-02179-9

The Point Press
Box 475
Cape May Point, NJ 08212

Cover and book design by Ann Giacalone D'Ippolito

*Dedicated to Bartolomeo and Lena Giacalone, my parents,
who helped me to make sense of my world, and
Christina, Lisa, Anita and Lewis, my children,
who helped me to continue the lessons.*

CONTENTS

It's in Their Nature

Right Answers

The Mind/Body Connection

Getting from Here to There

Practice and Performance

Personal Processes

Connections

Communicating

To Risk the Dream

Eyes and Ears

Sharing History

Aknowledgements

This book is no different from any other in its indebtedness to many. Meryl Nelson, whose patient collating and retyping of formerly published columns allowed me to desert *Pagemaker* and marry *Word*; her husband Bill, whose interested and interesting comments on the side caused me to rethink my idiosyncratic style; the members of our Cape May environs *Writing Group*, who listened and nurtured; Nicola Wynnefield, who kept her interest in my old stuff and also kept the original published columns; Wayne Richards, who cultivated my *resonant ear*; Sister Grace Madeleine, SSJ, who long years ago sent me on my first writing adventures which then appeared in print; my mother, who willed this book; my children, who, when I doubted, urged me to persist. Thank you all and thanks to the silent many whose support always amazes me.

It's in Their Nature

The seed already possesses the internal dynamic for growth.
It simply needs the careful gardener
to provide the environment.

The Seed Knows How To Grow

This spring I resolved to make a garden. Perhaps now that my children were grown, I needed to renew my familiarity with the beginnings of things from a different perspective. Also, I was lured by my vision of the finished project. I could imagine the results; where certain colored blooms would be, the placement of the table and chair for writing, where the birdbath would bubble. I read flower catalogues and some how-to-books.

Though the leaves were not yet on the trees, it was immediately clear that some plants required more sun than my oak bound yard could provide. I called in tree experts to prune and chop. Overgrown shrubs were moved. Tools, growth inducers and weed inhibitors were purchased. Some seeds I started indoors; other seeds and bulbs went right into the cold ground.

For a while, I was a daily watcher and hoer. Unfamiliar with many of the plants, I didn't know which end of the bulb was up, nor the difference between weed and selected seed, nor how to remove weeds without also taking up the barely rooted young flowers. But now, at this beginning of summer, some amounts of water, sun and fertilizer nurture the garden blooms. I guard less and enjoy more. This feels familiar.

Several months ago I asked a friend what she expected for her young children as they set off from home to go to school. Her reply, "I want them to know how to work within the structures of society outside their home, yet I want them to be able to think for themselves and not just take what's dished out to them." I was surprised at the completeness of the response and at its exclusion of such phrases as *good education*, *excellent curriculum*, *to excel*, *to make friends*.

Poets, patriots and other parents and pedagogues might color these same ideas differently, yet no matter how they are stated, the need to accomplish those two tasks she so simply phrased drives most of us all our lives. What an assignment for a young child barely into the first decade of life. No wonder parents and other educators become so easily confounded by the various and conflicting ways of nurturing our children.

The first experience of setting out plants, and setting out children, do have some things in common. Each certainly demands an investment of time, energy, a safe home and, yes, money. Each pulls us to be responsible for a new living thing that, no matter how dependent initially, has an inner dynamic quite separate from us.

Despite this, the nagging question persists, "Have I done enough?" We search out child experts; purchase all kinds of equipment to stimulate the child's physical, social, mental, artistic growth. We pluck out the childhood weeds that might choke that growth.

Someplace, sometime along the way, like success-ful gardeners, our experiences and instincts tell us when to let go and let be, and we exchange responsibility for response-ability. We listen to that separate dynamic in each child and take our cues from what we hear. Actually, at this point, we give our child permission to be and to es-teem that being. As the significant others in a young life, we are now able to nurture the self-esteem which allows the child to use himself well, either with others, or alone, as life requires. From us he then gets the strength to be himself and still function as a member of a group.

The seed already possesses the internal dynamic for growth. It needs the careful gardener to provide the environment for that growth. But don't over water. Didn't someone somewhere say, "More plants are killed by over watering than from being dry?" The art of parenting also requires knowing what's natural, what's too much, and what's enough. One just needs response-ability.

The Urge To Make a Mark

The first cave men began to keep a record of their lives by drawing on the walls of the caves they used for shelter. The urge to make a mark was so intense that cave people drew great moving figures in a place of total darkness.

Our preschoolers begin the same way. With paint brush in hand, and the large blank paper on the easel before them, each one begins to make his mark with lighted eyes, slow and careful application of paint, and intense attention. From cave people to people in the fastest century in history, the universal need to say, "I was here!" still persists.

Grunts and groans were not enough even for a caveman, so language followed representational figure drawing. Walls and stones were now holding symbols of the spoken word. Then, quite naturally, the need to know what someone else did, wrote, and said, produced readers.

I watch the youngest children mimic the reading adults in their lives. They flip through pages of the book they're holding and tell the story as they remember it or as they imagine it. The first word they actually read is their own name. When they finally manage to get their eyes, hand, and desire to work in harmony, they write the first letter of their name. There is one brown shingle on the back wall of the house where I was born, that still has an "A" carved into it. The need to make a mark and to know what one's peers have written produces readers as well as writers. There is a strong emotional element in reading and writing.

No matter that our society is such a fast moving, distracted one, our children are still fascinated with the hard work of making their mark and understanding someone else's. It is in their nature.

I repeat this little developmental story often to remind myself that the way I learned what I know, is far more important than what I learned.

As an educator, filled with the desire for all children to reach their potential, I can be overcome with the puffed-up notions that I can actually teach a child something. I slip into using knowledge out of context. I impart the information without the frame of reference. I begin to require drill work and testing out of my own need to measure what I have achieved. I lose contact with how these children can actually use what I know. I forget they are internally motivated to learn because they are fixed with that desire from birth.

Society worries us with terms and statistics. There is yet another word for parents and other educators to worry over these days. You remember when we were all deluged with "Why Johnny Can't Read" some years ago. Now the world's demand for a technological way of life has suppliers of such stuff concerned with not being able to find persons who can *use* mathematics. It seems that we still produce generations of children most of whom have only shopkeepers' levels of mathematical ability with little capacity for critical thinking and predicting. This feels like the old reading and writing story all over again.

If to grow and to learn are for most of us, part of the human condition, then as parents and other educators, we need to spend less time drilling and more energy on suppling the emotional and physical environment that demonstrates the uses of information. If I open the way for the child to be curious and to want to try, I open the way for learning, no matter what the specific field of study.

Rote learning out of context is stultifying, meaningless, non-motivating. In this day of computers, we need to spend less time with rote learning and more time in developing ways to find the problem-solving potentials in the world of the child. Find the context and the desire to master the material will follow. It's only human.

Surprise and Recognition

This morning I cannot ignore the very *springness* of the day. The bird songs follow me from room to room. The sun and all the yellow living things outdoors blaze brightness that breaks through each window. The outdoors comes in to lighten my tiredness and half-sick flu feelings.

Spring is a season of the soul. I don't need the calendar to alert me that it's here. Spring can even happen before the season officially begins. When it does, it carries with it an *ah-ha,* that exclamation of surprise and recognition all in one.

I was standing near a child the other day. He had taken the long route to clicking-in to reading and writing, so I wondered what he was so busy marking on his scraps of yellow paper. Paper was the one thing he had always avoided. Finally he tapped my arm and held the paper up to me. Unprompted by any teacher, he had been practicing writing, always his *bete noire.* At the bottom of the page, his name was not just printed but was written in cursive. He made the leap to spring

The spring in his face was matched only by the same glow in mine. There is nothing so reviving to a teacher as the *I did it!* look in a child's face.

The light in this child's eyes had not been extinguished by some over zealous mentor set on bringing him up to grade level. During his struggles with reading and writing, he had not been made to feel he had to be on everybody else's timetable. His teachers trusted his pace, kept the uses of reading and writing always in evidence, and then left him to do what he could. As a result, the child's love of being read to, and finally of reading to himself, remained intact, as did his self-esteem. So he pushed himself to practice. His own drive, rather than someone else's urging, allowed him to master a skill. The same urge as the one when he took his first steps alone and then wouldn't stop walking, takes over now too. It's as natural as spring.

I can have lots of springs when I refrain from teaching content, and remember, instead, that I teach children. In the first I fall in love with the subject matter; in the other, I am in love with the child, as he or she is.

Tell Me

As far back as I can remember I wanted to be told stories about how it was when my father was a boy in Sicily. Now I was sitting with my father in the backyard. With pencil and pad in hand, I asked him to tell me that story. Just like that! What a story blocker! I was full of the optimism he had always nurtured in me and believed that just by my asking, he could tell me. I have no other recollection of Papa sitting in the sun in the middle of the day, but today he was recuperating from a tonsillectomy.

I didn't write a word. The nature of a productive interview was not yet in my repertoire. He told me only one brief story. It remains with me to this day.

Papa's oldest brother was the family member usually sent to check on whether my father had been to school or if he had gone swimming instead. On the sunny Mediterranean island where he grew up, the water was a near temptation. His brother knew that a simple question wouldn't elicit a trustworthy answer, so he opted for the more direct approach. He licked the top of my father's ear to check for the salt that would say that swimming won out that day.

11

What could I do with this small story, when, in fact, I was looking for a book. Now, so many years later, I realize that this anecdote, juxtaposed with his life, was a more pervasive influence than I could have imagined on the sunny morning we sat together.

The young boy who dared to play truant managed to complete only three formal years of schooling. He let his imagination take him beyond those limits. He crossed the ocean and learned a foreign language. Daytime he worked and in the nighttime he studied and then passed his board exams. He seemed undauntable. He earned his profession and his place of business. He married. In all the years I knew him, his body was tied to the day and night hours of the pharmacy.

I always think of my father as the family risk taker, however. He would just catch up with one change in the pharmacy, when he would launch another: remodeling, enlarging, trying new public relations *traffic makers*. And his dreams took him beyond the drug store: a factory to produce fly repellent; donut machines; and a brief honeymoon dabbling with race horses. But the pharmacy was his constant, and he persisted. Who would have imagined such dedication to profession, family, and entrepreneurial pursuits from a boy who might lie about playing truant?

I believe that because of this story, I am a more tolerant parent and I am a more discerning, evidence-seeking teacher. I can be more relaxed at both. What I used unconsciously during the early years of my teaching and parenting, I can now trace back to his story, and the feeling of the book he lived during my lifetime. They play in harmony to remind me of the many ways we come to be who we are.

So often as parents we tend to use our child's behavior as a test for the quality of our parenting. I suppose many of us would believe we had failed to instill a love for learning, or, at least, a sense of reliability in our offspring, if our child dared to sneak off for a swim rather than go to school.

Wrong Move/Wrong Season

I inherited some misplaced tulip bulbs. They were growing on the side of the old house I bought. I thought they would really look better along the front of the house. Even though they were just about to bloom, I quickly transplanted them. Wrong move in the wrong season! When I looked at them later, the tulip buds were drooping on the ground. I took the unopened flowers inside, put them in water and hoped that they would revive there and not be entirely wasted.

I put the vase where I thought the flowers would look good and maybe get some sun. But the flowers were moved at the wrong time to the wrong place and so vital nurturing was withheld. Sun and water could not remedy what was lacking. Of course the flowers not only didn't open, but they also withered and died. This reminds me of another story of faulty caring.

One of the more quoted pieces of research on human development is the Spitz study of orphaned infants conducted after World War II. By now many of us know about the one hundred infants who were taken care of in every possible way: food, shelter, clothing and medical attention. The only thing that was withheld was affection. Caretakers were not permitted to hold the infants. Later Dr. Spitz remarked that he would not have conducted the study if he had foreseen the results. The infants simply did not survive. They withered and died. They were denied human touch.

Killing off some tulip bulbs and killing off a person, or that person's potential are certainly not the same. Yet each underlines the same kind of wrong-headedness. Sometimes we can get caught up in looking good. We hurry, or ignore timing, because our wants don't coincide with another's needs.

Using a child, a spouse, a friend, a work to make one look good, is enough to kill off the kind of care that allows development. They and we begin to exist primarily to fill an external need or to create a front. Love changes and becomes a bribe to perpetuate the appearance of caring. Caring becomes *caring for* rather than *caring about*. Love is reduced to filling physical needs: food, clothing, medical attention, even good schools and social events.

The work, the family, the child, become artificial and stagnant. Then love for the work or the family or the child dies as each loses its specific reason for being. A child, in fact any human of whatever age, needs to know that what he feels, what he struggles to understand, and what he loves, is worth something to the significant others in his life.

When I talk to a child, I need to be able to do more than ask questions or give directions, or impart some wonderful morsel of information. I must stand right where the child stands. I need to forget what any other child is doing, what any other parent is saying, or what any other relative is expecting. I will not judge him or remove my love from him for any reason, not in the name of caring for, and not with any intention of impressing anyone with my way of parenting.

I talked to a former colleague last week. We talked about the ways in which humans can provide a *resonant ear* for each other. He meant, I believe, that if one listens to another well enough, the person speaking would be able to hear himself. The listener becomes a sounding board. He plays back the story and enhances it just by the way he listens. So the resonant ear says, "I care enough about you to put myself in your words, your thoughts, and your feelings."

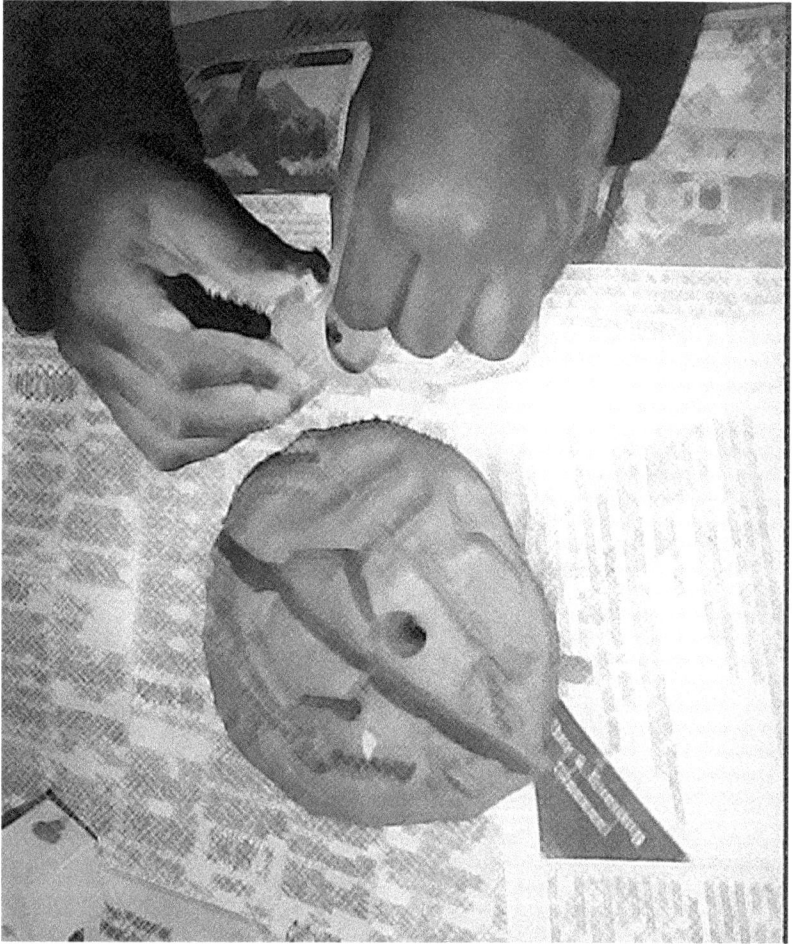

Right Answers

Master teachers discover talent and nurture it.
They do not invent it

.

Wacky Preconceptions

A new staff member shared an anecdote with me this week. From the number of times I've already retold it, her story seems destined to become one of my all-time favorites.

It went like this. A teacher asked her second grade class, "Why can't you hold a cello under your chin?" Bright-eyed and sure, a young student volunteered, "Because you'd get a hole in your neck." "Wrong!" corrected the teacher, "because it is too big."

Doesn't this exchange sound as though it came out of a Ritz Brother's comedy? What's not so funny, and you can tell this by the seriousness of the teacher, is that it wasn't intended to be comedy. Take one very serious, there's–only-one-right-answer teacher, and mix with the uncluttered, unfettered minds of second graders and we have a dilemma in chemistry. What will happen to the two elements? Will they become a compound a mixture, an emulsion, a volatile gas, or ... or ... or ?

Some children grow up believing that there is only one answer to any given question. They come to feel increasingly secure in proportion to the number of right answers. And this brings another story to mind.

I met an older couple years ago. They had been married for a long, long time, and had known each other since they were both wealthy, young socialites. They had remained together even after they lost their fortunes. When I knew them, he seemed the patient, caring, and solicitous one, who had turned whatever abilities he always had, to growing orchids in a nursery. He made a living, took care of much for her, even made sure she was entertained with car rides on his days off.

She, on the other had, never understood her life once the money was gone. "You see," she would say, "I never learned how to do these things. I was treated like a doll. I don't know what to do with myself. I'm afraid. I don't know what's going to happen tomorrow." She had learned only one set of answers to her life.

When factories proliferated at the beginning of the twentieth century, schools had the job of taking emigrants' children from diverse foreign countries and making them the same. At that time schools were built like factories and children carried the same lunch pails and answered to the same whistles at the same times as factory workers did. As much as they were able, all children learned the same sets of answers to the same sets of questions. The job was successful and emigrants were homogenized.

The industrial revolution is passed, but we still get more excited about products than process. Products are consumable. Process teaches us how. The first is quick to make but is quickly gone; the second takes time and lasts a lifetime. If we don't want to keep our children in a mental state that is equivalent to a conveyor belt, we need to ask questions such as: *How did you get to know that? What did you have to do to understand that? How did you learn how to do that?*

If we want children who can think, we have to share the rehearsals with them. We need to show them how we learn to do our work. Then they must practice for themselves. If they see us engrossed in our work, they will want to have a work of their own.

The learning process is a story of personal growth. People, like other living things, grow from the inside out. As parents, and as life learners, we need to combat the fear of the struggle, the tentativeness of the mystery, and the lost feelings in the face of the unknown. With the gifts of time, patience, and respect, growth will happen. Then we can stop alternating between the roles of performer and audience as we interact with our children. We can begin to participate in our own and each other's lives. Learning is not an option. It's a survival skill.

Muscle Bound Minds.

Some people have muscle bound minds. It's not so much because they don't use them, or that they use them too much; it's that they use them one way, excessively. What develops is what a friend of mine calls, *musculature of the mind.*

The varieties of exercises that promote mind petrifaction are probably endless, but I have some special non-favorites. At the top of the list is *either … or ….* Surveys are great for over exercising this particular brain muscle. You know, the kinds that ask, "Are you a conservative or a liberal?" "Are you traditional or permissive?" "Do you consider yourself a success or a failure?" "Do you support public or private education?" I'm always prompted to return the favor of the questions: conservative compared to what, liberal contrasted with whom, permissive when it comes to what issues, success at what, failure when, which private or public schools?

True or false questions, and the assumption that there is only one correct answer to a given question are other inflexibilities that straitjacket the mind. Slogans also can have the same effect. *The best defense is a strong offense*, infers that the only successful response is an aggressive one. To motivate, slogans also present either-or scenarios and they close the mind to the variety of possibilities inherent in every situation.

I'm bored is a classic example of closed, muscle-bound thinking. Where would a child learn that word? Children learn to close off or to open themselves to new ideas and experiences by what the people around them model. I have a personal theory that children who have been deprived of enough time outdoors or with animals have a greater tendency to fall heir to the bored trap. No child who has been brought up to appreciate flowers or birds or fields or the sea or the change of seasons or even the extraordinary changes evident in each ordinary day, will ever say, "I'm bored." In fact the boredom trap is not outside the person, it's within.

If by labels, slogans or either-or solutions, I restrict the way I view a person, a topic, or a situation, I use part of my brain repeatedly. Then I run the risk of a kind of mental hardening. When I can be as open as nature, when I can see the variety and complexity in the smallest grain of sand, drop of water, blade of grass, when I can be awed by the immensity of infinite space, then I begin to understand the world that surrounds me.

When I can look one more time, listen a little harder, or talk once more even though the situation seems impossible, I begin the knowing that allows for real communication with the humans who walk this world with me. Then I can begin to be a teacher.

She Has No Choice

"She has to learn to conform. When she's an adult she can decide whether or not she wants to be a non-conformist. That's not for children. Right now she has no choice."

"Conform to what," I wanted to ask this examiner, "to who's what and for what?"

She sat efficiently marking off the items on her sociological checklists. She said she had never met a child who didn't respond to her. I was not surprised that this child had not. She must have sensed the rigidity in the older person that was the enemy of her still fragile creative spirit.

I don't really believe this adult had two standards: one for adults, and one for children. In fact, I sensed the adult was as rigid with herself as she advocated being with children. I have a hard time picturing her playing with anything, and yet she talked about being able to take a risk as part of the equipment one needs as an adult. Conforming and risk-taking don't seem to go together.

The other day I caught myself wondering aloud to a parent as we were trying to puzzle out her child's learning style. "He seems to learn everything peripherally. Maybe that's the clue, to have him overhear what he needs to learn, learn out of the corner of his eye (or ear), so to speak. Maybe we're making a mistake in trying to get him to focus straight on."

Teaching and learning, whether one is in the classroom or at home, seem to be interchangeable. One cannot teach, and teaching cannot be confused with training, if one does not take the stance of the learner. Not just a pretend stance that feigns not knowing to seduce the student into giving an answer, but a genuine belief that no one has the answers to another's learning style. To know the student, one needs to allow the student to take the lead. Master teachers can do this. Master teachers discover talent and nurture it. They do not invent it.

Conformity would make us all the same. Yet one knows beyond a doubt that everyone isn't the same. Some of us need to touch to understand, some need to visualize, and some need to hear. Some children need to call reading "cooking," or "washing clothes," or even "math," anything but what the adults in their lives want to title "reading."

Since reading may be a weak spot, and can be perceived as a threat to self-esteem, a child experiencing difficulty will try to avoid whatever is called reading. If a child still has a weakness in reading that has been taught conventionally, maybe he's telling us that our convention isn't his. For a while, until he's more sure of himself, he can't take reading straight on, but he can get to do it out of the corner of his eyes.

27

What do we do when we ask a child to conform to a standard that underlines his weakness? What do we say about the child's worth? It seems most important that the self-esteem muscle is built before a parent or other teacher can hope to strengthen other muscles. Self-esteem has never been nurtured by demanding conformity, that we all do the same things the same way in the same time frames so that we can all be alike. Risk-taking as an adult has never come from a childhood that places primary importance on doing things like everyone else.

My observations become blurred when I judge a child, or any one else, by what I think he ought to be about. When I do, I can't see the child's present need or his future possibilities. Perhaps I even confuse his ability to see himself clearly. What then becomes of the quality of his dream, or even his ability to dream?

Dreams seldom come from a conveyor belt life. Dreams are individual and they are nurtured as a person senses that his unique history is unduplicatable

For a Lifetime

I need a condo. Maybe what I really want is a small house. An asset in one year seems a liability in another. Often financial answers seem as elusive as the *right education*. I am driven to ask, "What does endure? What maintains its value for a lifetime? What is worth passing on?"

The last years of the twentieth century were full of nostalgia journeys. We wanted to hold on to what felt safe and beautiful and valuable in another time. Nostalgia can be addictive, however, and looking ahead at the unknown, unfamiliar, and changing future is frightening. The road ahead is pretty dark.

When I look at our children, the answers become even more difficult. Our five year olds will be fifteen in ten years. They will be looking at themselves and asking questions about what they need for what they *want to be*. Our ten year olds will be twenty in ten years. They will wonder if they will have a job and what that work will be. No matter what century, this means having not only an identity reinforced by some innate talent, but it also means having a voice.

Identity and voice are not formed at fifteen or twenty. They're nurtured, or not, at the very beginning of life. Significant caregivers in the early years are the esteem and confidence builders of the future adult. Esteem and confidence allow the child to recognize and to use talents, those talents that are unique in each of us.

Parents and educators can provide openings for talent to emerge. Choices within age appropriate parameters show the child that she can be herself and be okay. This gives the emotional self the permission to try, to handle mistakes, to know that setbacks are temporary and can be strengthening. The child can begin to feel that life's changes are manageable. On the other hand, the fear of mistakes and not measuring up generates people who are afraid to try new things. For these fragile persons change becomes frightening and everything unfamiliar becomes untrustworthy.

Facing ambiguity, making mistakes, and adapting to what is different, are important experiences for all of us. Like a child learning to walk or do stairs, surmounting obstacles teaches me to *learn how I learn*. Then I understand how I use myself as a reliable point of reference for dealing with life.

Contrary to some myths in our culture, the scientific method, which some say provides answers, is really based on asking questions. A significant question can become much more important than someone else's right answer. The child, and the learner in all of us, needs to have the experience of wondering, of being comfortable with the blank page, or of sticking with a question until a variety of ways of answering are found. This is how we learn how to be discoverers. The unknown becomes an adventure.

Openness to the unknown and to variety is an indispensable exercise for learning how to use myself as an authority. How different this is from feeling incapable of making a choice or of always needing to be told by the expert. As I make choices I can now utilize the value in what is said by others. But now I use myself as a sounding board.

Contrast this questioning openness with the pressure of learning to satisfy artificial standards. In the face of arbitrary standards I need to ask, "Which standards for which tasks?" With all the information available to us, how do we know what is most valuable? Can we shortchange the options we provide to our variously talented children? To which of the multiple intelligences do we appeal? In this process, don't we favor some kinds of intelligence over others? What does this do to the child's ability to know and to value what s/he actually is? This is when we say we hold the answer for obtaining the prize, and learning becomes a contest.

The natural excitement of learning is different from the artificial tension of a contest. One raises the instructional level; the other produces a frustration level. Learning is short-circuited. The first gives the child a voice; the other robs it. In the second we imply, "What you say is not worth listening to."

When I know a person's style of learning, I open the door for all learning to be discovery of the individual voice she possesses. To know my voice is a value that lasts a lifetime and beyond.

31

The Mind/Body Connection

I wonder if our children know
that being always-on-the-go is not necessarily
knowing where we're going.

Paying Attention

A friend of mine gauges the state of his mind and body by the condition of his car. If the gas is low, the oil needs to be changed, and the paint hasn't been washed, he's sure that he is also going around with a near empty wallet, too little sleep and not enough nourishment. What would he need to do, he would wonder, to keep track of both car and body before depletion set in?

Generally, I feel that I do pay attention, but I notice that my car hasn't had a wash in months, the oil certainly needs to be changed, and I am consistently rolling into gas stations on fumes. I've been borrowing dollars, there are circles under my eyes, and I have not had the right kind of food in the refrigerator to help reduce my elevated cholesterol count. Someplace I stopped paying attention.

Somewhere, in the frenzy of getting ready for new seasons, new demands, and transitions from one phase to the next, I begin to take on an exaggerated sense of what I can do. When I get to this point, I notice that I begin to feel less able, not more capable. I make mistakes, forget and need more energy to finish each task. Lack of exercise gets me to feel *exercised*. My frustration threshold lowers. One more demand will put me on empty.

Ricardo Muti conducted the Philadelphia Orchestra for several seasons. He commuted from Italy to Philadelphia to keep this commitment. He decided to give up his post. His reason for quitting: "I want more time to spend with my family." Peter Lynch, formerly of Fidelity's Magellan Fund, whose income had been assessed at ten million dollars a year, retired at forty-six. He wanted to spend more time with his family. "I am missing so much of it. I went to a soccer game with one of my daughters, and I had a great time. I went to one soccer game; I missed seven."

Great, but I also know when just one more soccer game will be one too many, and that's the other side of it. Overload can come with too much work and/or a frantic attempt not to miss anything. I would have liked to talk to Peter Lynch, or Muti, for that matter, to have them describe what being with the family meant to them. Some of us suffer from too much involvement with each other's lives, and some of us from too little.

What is the balance between wanting to wear out rather than rust out, and being all worn-out; between missing so much and wanting to be able to miss a few. I know that running to endless classes and lessons and cultural and social affairs does not insure that I am not missing something with my family. I wonder if our children know that being always-on-the-go is not necessarily knowing where we're going.

I'm not sure I have the answers to the polarizing questions life seems to pose. I sense that if I can make just enough quiet time each day, I can find out what I am doing or not doing and what I need to let go of. If I am present to myself, I can be more than a stand-in with my family, friends, business associates, and colleagues. After all, I don't really want the people I value to miss me more when I am there than when I'm away.

After weeks of feebleness, the sunrise held promise this morning. Its beginning was pale, but now, true to its first promise it burst forth in an almost white brightness. Little reason to wonder why the ancients chose this springtime of year to celebrate rebirth and renewal!

Time Expands

There is a small window right under the eaves of this slanted room. As small as it is, it frames the view perfectly. In the foreground I see the lovely garden below; in the background, the border of dune grasses outlined by weather fences. Finally there is the sea beyond. Though ships pass, fog rolls in, the sun sets, and the moon rises outside it, this window seems to stop time. It's a vacation window.

When I'm here, I don't need to answer phones, keep appointments, schedule my hours and minutes, nor cram a lot into too little time. Here, almost magically, I can feel time expand. There seems to be more of it. The bird songs sound slower, even the rabbits are more deliberate. In a world that always feels as though it's moving too quickly, I want to know how to find this slowdown when I'm not on vacation.

For a long time, I've talked about being able to put on a vacation mentality no matter where I was. Daily I would look for a way to drop out for a while and to imagine that there were no external demands and that I could respond to my own internal clock. I've become really adept at using music, writing, poetry, art, and meditation to get me to this other place. I had become so good at this exercise that I was convinced that I could go years without taking a real away vacation. Now as I look through this window, I come to understand the wisdom in taking a vacation. There is a time when a vacation also needs to mean a place as well as a state of mind. There are times when one needs to get away to be away.

A short walk from my window is the beach. For the last few days, I have been entertained by the seemingly mindless yet purposeful blend of nature's summer rhythms. I have watched headless horseshoe crabs submit to the push, pull and abandonment of the ocean; small almost transparent crabs climb quickly into the bottomless holes that dot the edge of the water; and finally, vast numbers of porpoises, literally porpoises by the mile, do their rhythmic dances along the coastline. But now an old familiar pattern comes into view and now even the porpoises cannot compete for my attention.

I am drawn to watch a father and son's progress down the beach. Each one appears intent on observing as he walks. First the father stops and bends over. The son does the same in another place. Then they bend over side by side. They are clearly doing this beach thing together, and yet what captures my attention is the freedom apparent in their movements. No coercion; no manipulation. They seem to have a rhythm undirected by anything other than the pull of their separate wonderings and wanderings, yet they are in time with each other. They alternate leading. First one points to his find and they both examine it, then they go off on more individual hunts. As though by some secret signal, they come together again to

39

exchange discoveries. I am captivated by how wordlessly in synch they are.

This vacation spot is new for me. Maybe it's this newness that has a way of making even the most ordinary things seem fresh. What I see, hear and feel takes on a new significance. Some of my old patterns are becoming more visible in this different light.

The press of my routine existence requires that I drop out from time to time to de-stress but the dynamic of this different vacation space seems just the opposite. Here I drop out to drop in. I leave familiar surroundings to see afresh what had become so familiar that it had become invisible.

If the extra breath of time that vacations bring does anything, it allows me to discover my internal rhythm. That accomplished, I can, like the father and son team, come to lead, and to be led, to wonder and to wander away, to know my own timing and to know the timing of those whose histories I share. Relationships are nurtured without the press of time.

I've Begun To Make Lists Again.

The computer, rather than the sun, casts its glow on my face. The phone rings off the hook. My allotted vacation time has elapsed. I put summer away reluctantly.

I will miss the longer stretches of free time, time to play, and time to watch the outdoors closely. I know from experience that when I lose contact with this physical world, I also miss the cues my body sends me. Then I begin to function too exclusively out of my head.

Maria Montessori, the scientist most famous as an educator, made her commentary over a century ago about society. She said our world consisted of persons working as either minds without hands, or hands without minds. We haven't changed much since then. It's true, we have become a more fitness-oriented society since those days, yet we still seem to miss valuing what the physical world can bring to the way we think. We still separate the physical from the mental and the mental from the physical.

I believe that exercise and working with my hands are valuable to aid my thinking. My behavior, however, often says that I give my mental tasks overtime, while I

41

promise myself that I can catch up on exercise and play *later*. I can't and I usually do not.

Are my childhood school experiences responsible for this split in the way I've lived my adult years? I was fed a steady diet of physical education. But is the catch allotted time? Physical exercise and recreation were not naturally in the flow of each day. Fun times were not spontaneous. Play came after. "Do your work first, then play," my mother always stressed. The physical component was for a certain time and place. Play came after all tasks were completed.

Most of us understand what it takes to be creative. We just don't believe it's for us. We have to work. But if work is to take us to another place, to find something new, we need to have creative insight. To be creative is to be able to play. To play is to move.

Even now, our expectations are for our children to sit at desks and to use only their minds to complete their tasks. Since the turn of the twentieth century, we have recognized that humans don't work best that way. But for most of us, beliefs and expectations don't match.

And how about this? In the last few months, I have heard people talk about what is ideal and what is real, as though the two could not coexist. I try to discern what they mean. Do they mean that education that is ideal is not useful in the real world? I've thought about this for a while, and ask another question. Is education preparing us for jobs in which we must fit a certain mold to succeed? Is this a repeat of the education pattern of the industrial revolution? Is this a way to keep people in their place?

Following what someone else has devised is just that, following. Dreaming about what can be develops leaders in every field. And this often requires separation from the group. One needs to step out of one's place to lead.

Do we provide a place for our children to taste how varied our paths to understanding must be? I can learn reading through math. Music can teach me to appreciate poetry. Car building can teach me science. Sports can teach me physics. I can often use the work of my hands to free my mind to learn.

Doesn't it follow that if a child can use the area in which he has the most interest and talent to pursue what is more difficult for him to learn, s/he can become excited about learning? As a parent, as a teacher, when I allow this to happen, I provide the ideal teachable moment. I free the child to follow his inner dream, his individual talent. The child can understand leading himself and leading another.

I'm tired and I still need to think. Let's go for a walk on the beach to prove the theory one more time.

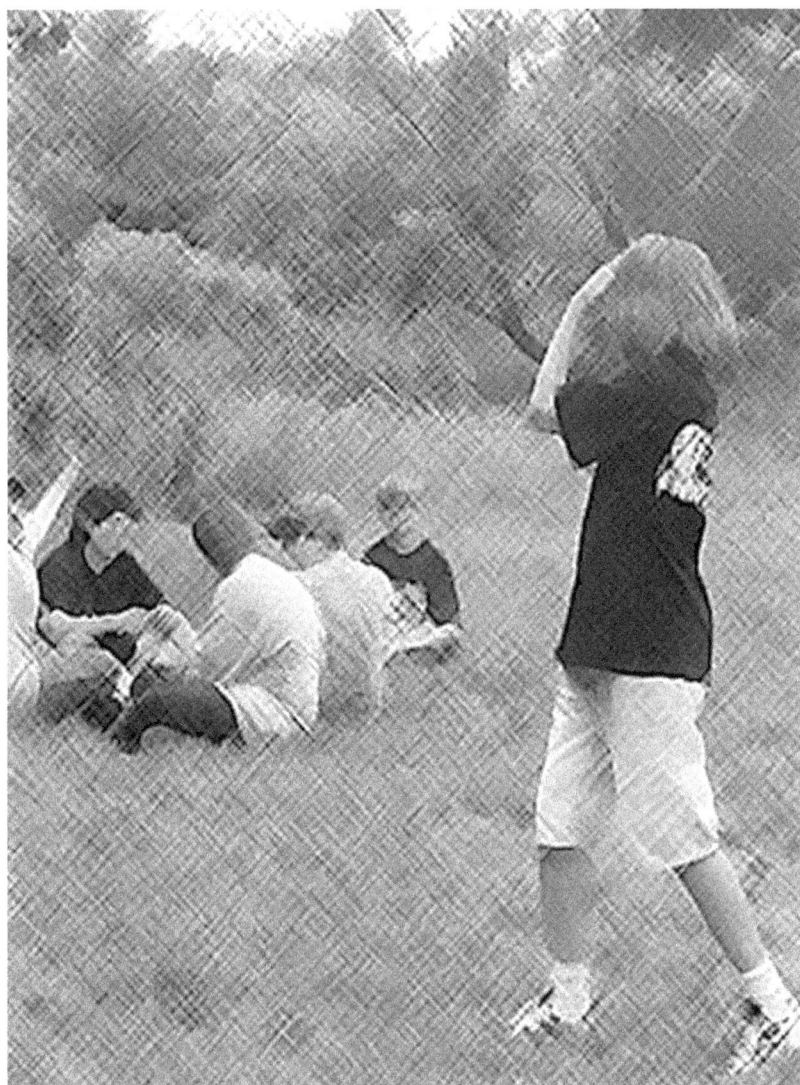

Getting from Here to There

You seem to sense you're on the way to someplace,
even as you doubt, because on the way is never really there.
So you appear to drift between places, and
I wonder if I have lost touch with you.

Just Enough

Sitting on the rock overlooking the ocean, I admired the stones, just enough of them to fit in one hand. Luminous, subtly varied in color, size, and shape, they had all come from the same small section of beach. I don't know what makes one day's walk a time for gathering stones and shells, and another day a time just to inhale the sea and sky. This was definitely a gathering day.

As I moved the stones in my palm so that I could examine each one, a familiar rise and fall of voices came from behind me. Even before I could make out the words, the cadences told me the interactions were involved, non-threatening, and even friendly. The sounds came closer to the rock I sat on, and then finally passed into the arc of my vision, and I understood why my memory was jogged. How many times I had walked the beaches with my young children, very much the same way this father was *doing the beach* with his young barefooted, pony-tailed daughters. The sounds and movements were so familar.

I seem to need to pay attention to these beach walk stories, and to gather them up like the stones I hold in my hand so that I can see them more closely. There are few places that allow the same meandering interplay of people as the beach does. One is compelled by the ocean on one side, and the difficult too soft sand on the other, to follow a certain path. Yet the berth is wide, safe and allowing. It's easy on the beach. So now I watch closely as their walk away from me shows me who is the courageous one, who is curious but cautious, who collects, and who simply investigates.

I suppose my gathering of beach stories is restatement of a dream where our children are free, interested, and able to make choices, able to be themselves and where parents, can enjoy their children as those processes evolve. There is, after all, no greater gift one can give a child then the freedom to be himself. Yet I still call this a dream. Too often the way we fantasize being with our child gets lost as the dailyness of living obscures the ideal. Is it the *doing* that becomes the trap that prevents *being*? If so, perhaps I need to examine what I do.

Some things I do know. As a parent and an educator, I am obliged to understand how to make environments that are conducive to the kinds of things that happen on a beach walk: discovery, mutuality, separateness. I know that human development need not be a forced march, but rather a joyful celebration of the best uses of self and others as people, not as commodities. To do this I need to believe that every person carries the purpose of his life within him, and that I can only admire that inner directive. If I'm really a teacher, maybe I'll even be able to draw that inner voice along. Perhaps if I can provide a wide, safe, yet allowing path, very like a beach walk, I have begun well. The rest we can learn together.

The Blur

My daughter showed a new friend some of the streets where we lived when she was very young. Within the same block, her grandmother and great grandmother had their homes. She pointed out the houses of neighborhood friends, went down one block to the school and church she attended, down another two blocks to the *main drag* that held memories of shopping expeditions with siblings. Even the movies were only two blocks away. For the first years of her life, all that she needed was just a short walk away.

I have been on a nostalgia trip for years, searching for places that still have what I call a village life. I too remember the walking towns we used to have before cities large and small erupted into a series of sprawled out highway drive-ins of commerce. Now most of us have to drive or be driven to get to our daily activities.

I don't imagine that what I miss is just the walking and the biking to places that were part of my childhood. I do miss seeing more closely the persons who are part of this place where I live, and I don't like trying to figure out whom I know in the passing car so I can wave at the blur of that person.

I feel so many of us now have to rush to wait. We are usually bundling some child off to some place for some thing. Adult and child seemed locked together more than we need to be, simply out of our need to share the same transportation. Somehow in the process, we lose the rhythm of our separate lives, and even the rhythm of the seasons. Most of all, our sense of time feels constantly constrained. There is seldom enough.

Though all of us can do mental flashbacks to *how it was,* we can hardly reconstruct old village life (current developers of commercial "Main Streets" notwithstanding). I believe, however, that we are beginning to understand the need to pull back from the time race. Our children, most especially, need to be less rushed, less pushed, and given more time to be in synch with their own rhythms.

Maybe the issue is what it takes to lead, and therefore, to follow. The old saying that good leaders are also good followers is still true, as the best leaders know when to let others take the lead. As a culture, we seem hardpressed to find leaders. Leadership requires the ability to be so in tune with the self that one can follow without losing that self. What can we give children so that this passage to leadership can happen?

Children – and all learners – need uncluttered time to develop the information about themselves that can be their center of reference. We don't need more fast drives through our lives. We do need calm, slow walks through the personal village we carry with us.

Twelve Is Another Hard Number

You find it hard to hold the broom and your unfocused hands and arms scatter the crumbs to the four corners of the room. You seem simultaneously tired, excited, jaded (you call it *bored*), scared, happy, unsettled and truly miserable. Your front tries to say you're confident as you boast your way through these days. You joke a lot. I ask you if you know what gallows humor is?

You seem to sense you're on the way to someplace, even as you doubt, because on the way is never really there. So you appear to drift between places, and I wonder if I have lost touch with you. I took this road so many years ago. It's not easy being twelve.

Maybe I need to tell you that I too hear myself joking more than usual at some times, the times I feel the most uneasy. I too have those in-between times when life seems unsure and I can't find the signs along the way. After all, no one ever told me that I needed to plan for what I wanted to be when I got old. So until the last decade, I hadn't thought about a plan for aging any more than you are thinking about a plan for your adolescent years.

I wish we had more rituals to get us through these transition times. Rituals come from those who have taken the journey before. They're ceremonies that dramatize what is ahead. If you were to take part in a ritual right now, it would be one that tests your physical strength, but most of all, it would check your endurance. What can you take? But always it would be measured so that it would not be impossible, and you would have the elders of the community to guide you through the trials. You would know from them that you would be all right. They went through it too.

That's what a ritual does. It helps you to go from the pain of the present place to the new place. It tells you others have gone before you and that they survived, and even thrived. Those others tell you what you'll need to get where you're going to go. Then they give you a visible mark that lets everyone who meets you know just where you are. Rites of passage they've been called.

Cults have thrived in the recent past and still do. Maybe it's because they provide the rites of passage our culture has forgotten. But most of these are negative rites that lead to destruction.

I'm not sure adults now know how to make the connection with you that might act as a rite of passage. Outside of brief religious rites, our culture doesn't provide you with that help. My age and stage doesn't have that kind of marker either, so I do know something of your confusion. Marking the draft age, the drinking and driving age, the voting age and the social security age, just don't do it as rites of passage. Someplace along the path to this twenty-first century we lost the rituals that would prepare us for different ages and stages in a way that would enhance the human race, make us better, more able.

Maybe it's up to you and me to post some markers for those who will follow us – take on the job of being scouts that leave signs along the trail to reassure those who come after us. Maybe if I share some old stories with you and you share some fears and dreams with me, we can realize that we're not doing such different things after all. Then neither of us will feel quite so alone, or different, or isolated, and we can leave behind something that others can use. We'll know the differences among being poorly used, used up, and used well. We will be better, and those who follow us will not inherit a cast-off, used-up world.

Practice and Performance

The way I learn best is by watching someone else
who loves what he is doing, who allows me
to practice alongside him, and who can
celebrate with me as I finally enthusiastically say,
"I did it!"

.

Immigrants

When my father immigrated in the early 1900s, he settled first in Jersey City. There, he and his brothers and sisters went to work in a factory. My uncle, who came from Italy as a civil engineer, was the first to begin to study. Through some process of shared expectations, he decided to be a doctor and convinced my father to explore pharmacy. He probably thought the combination would be profitable. My father had only finished third grade in Italy.

Almost another generation before this, my great uncle, a very small young man, came to this country with the last immigrant group on my mother's side of the family, all women. He was so small that he passed for a younger child and was allowed to stay in the section on board restricted to women. Size and his immigrant status didn't keep him from being ambitious. He decided to study law. His enthusiasm and his professional offices later intrigued his nephews with the possibilities the law would allow them. These uncles also became lawyers.

In due course, my father's brother became a surgeon, my father a pharmacist, my great uncle and one of his nephews prosecutors, and the other a judge. They learned their professions largely from apprenticeships and private study. They interned and clerked until they knew enough to take their board examinations. They each came to excel in their chosen professions.

There are individual styles of learning, and there is also the optimum time to learn something. Two things are sure. One learns from watching someone else and one learns from practicing.

A noted physician on the board of George Washington University complained that the college's biggest problem was students who had no desire to learn. He questioned, "How do you get them to want to learn?" Nationwide the unrecognized problem is the inability to allow students to possess their own learning process. Today the pressure to turn out motivated learners seems to be doing just the opposite. In pushing quantity and competition, we forget that personal interest is the life-long motivator. Our pay back is students who become turned off to learning and who are only turned on by the carrot of success. Dropout, burnout stress-related physical ailments and high suicide rates amongst young people are the result.

We need to object. This is not what we want for our children and we shouldn't be content if this is our only choice. We don't have to say that feeling that bad about self or life is *the way it is*.

The way I learn best is by watching someone else who loves what he is doing, who allows me to practice alongside him, and who can celebrate with me as I finally enthusiastically say, "I did it!"

Coming into Focus

I can still picture the many intimate and long hours - just myself and the piano. I learned how to pull my thumb way under the palm to advance up or down the keys, and how to vary the pressure of my fingers to get more speed. I learned how sensitive to a variety of touches the eighty-eight keys, the strings, the hammers, and the resonating wood could be. I learned what the instrument could do, and I learned what I could do. I learned how to practice.

Yesterday I looked at a kindergarten child's folder. The first paper contained a printed page of "A's." Though the teacher had printed a guide for the formation of the letters and for their spacing at the top, the young scholar's "A's" drifted randomly across the page. As I got to the end of the folder, I found a page of "G's." I put the page of A's and the page of G's side-by-side. Looking from one to the other I got the image of a child coming into focus like the lens of a camera. Unlike the drifting "A's", the "G's" were placed deliberately on the page. The practice over the school term had brought this child to an understanding of the organization of the page. He now had control of his paper and pencil instruments. He had refined his execution of the task. The teacher had made no correction marks on either page. The child had been given the chance to observe and practice.

I walk through the classrooms of very young children and the walk becomes a lesson in human development. Preschool children at easels explore the materials. Some youngest children dab paint only at eye level, others paint the whole page, others use the reverse end of the brush to etch, others feel the paint with their hands, and still others swirl the paint over the entire large paper. The next age group I observe begins to show the emergence of a shape, a large circular form. Smaller, more deliberately placed objects begin to appear on the paintings of the older children. Placed in sequence alongside each other these paintings write the book on early human development.

I think of how many times I've practiced. Not surprising. Practice seems to be the way one learns everything. Practice is just that, practice. The first steps a baby takes involve lots of falls, lots of bruises. We adults who watch give assurances to the child that he will indeed walk. We know that the actual walk, like everything else we learned, is something between our bodies and the skill. One needs to experience for oneself the materials of the task and the body's response to those materials. So we leave the child to his practice.

No matter what the endeavor, a child needs to have the space to try, to stumble, maybe even to fail. With encouragement from us, he learns that each practice is not a perfect performance. Together we can delight in the explorations, and trust in his processes, which, indeed, may be very different from ours. If the child can feel comfortable enough to practice, eventually, to play, he will find his way to a personal excellence that can affirm his special talent.

We each master skills by some inner clock. Each timepiece is set differently. Reading, writing, computing all have timing that is internal. We do not move in lock-step in these skills any more than we all walk at exactly the same time, or get our teeth in the same month. We need to remember this in relation to academic skills. Remember how we learn to walk. Trust the child's timing with tasks that need to be practiced.

For Public Consumption

This is one of the performance seasons. Weddings, graduations, end-of-the-year school productions, each focus the public's attention. Even our front lawn measures how our public face holds up. What kind of a product do I produce? Most times I can absorb my blunders when they are private. When the public is asked to witness an event, the possibility of a mistake can strike terror into the most blasé heart. After all, *What if . . . ?*

There is a time, no matter how private a person I am, when I do have to share what I know. Sometimes knowing is not enough if I can't communicate that knowledge. This requires some practice in performing. Performance is often accompanied by anxiety. Professional performers get some emotional payback from performing, yet I haven't met one who doesn't go through the same angst of *What if ...?* before each performance.

Last week our children practiced for the school's ethnic festival. It's when I watch the child working through the practice/performance pattern, that I get a vivid sense of what it means to practice, and how it feels to perform. During practice, children have a hard time going through from beginning to end without stopping and redoing the glitches. Eventually they get the routine down pretty well but then it becomes so automatic that they stop thinking and the glitches reappear. Finally they get to performance time. The teacher says, "Do your best and no matter what happens, just keep going."

I watched the children. They had pride in what they could do. They understood they each had a separate talent. Someone else's way of being or doing didn't distract. There was no time to covet another child's talent. They each had a time to shine and they each had a time to perform with the group. I could almost feel the child saying, " I can do this performance thing without being afraid of losing myself."

After some weeks of close study of an ethnic group's food, clothes, music, art and history, each class of children from toddlers to seventh graders performed a dramatized vignette of the ethnic group they had chosen. I watched as they did their group thing and then as the individual child contributed singly. After the performance a parent said, "You know, this was great because it allowed each child to act individually and at the same time know that everyone else would have a turn, and that the group was there right behind him."

It's good to have this first experience early on with group support, a caring family, and a nurturing school. When this happens a child can recognize that the strength of the group depends on all the differences joining together in one voice.

Enough?

Last week an artist I know was invited to a special New York party. Her first commissioned painting was being hung. She spent the time before the event wondering, and soon the excitement over the commission being completed and hung was overshadowed by questions of: *Will they really like it; is it as good as I thought it was; will everyone respond to what I hope the painting says; can I go to this party and hear all this?*

No matter what career one chooses, there is that point at which one wonders whether the results of hard work, talent and dedication to an idea and an ideal will really bring what one expects, what one wishes for, and what one hopes for.

Positive response to one's work says, "I have been listened to and I am affirmed." Negative response says either, "I can't hear you," or "I hear you and I give nothing in return." Elation and disappointment are two sides of the same coin.

Holidays seem to underline the quick flip of that coin. Expectations for them involve so many levels of feeling and doing and are wrapped so tightly with the myths of our culture about what a holiday is supposed to be. Children, in particular, have a hard time fitting the intense excitement of the weeks before the celebration with their feelings when the holiday actually arrives. They begin to question if they have missed something. Parents can be disappointed too. "I planned all this and my child didn't really respond enough, wasn't surprised enough, wasn't excited enough, or disregarded what I gave, preferred something else." Are the holidays to represent the foreshadowing of things to come?

Parents often need to see through their children that they have done a good job. I suspect that one needs affirmation for any work. Our expectations around parenting are no different from the patterns we set up everywhere in our lives. The question then becomes, "What is the measure of any job? How can I teach this to my children?" The end of the party story seems to answer both questions.

> *As I went into the loft and took off my coat, I gave a quick look out of the corner of my eye to where the painting hung. "It's good." I thought.*

We learn to answer many *if's*. If I can recognize my personal best, if I can weigh what those around me offer, if I can pay attention to my patterns of response to others, then I can begin to trust my talents, my integrity, and my personal best to be enough. If, as a parent, I can demonstrate this to my children, then they will have a model to follow. They will learn to understand that they are enough.

Personal Processes

Special is individual. It is patient,
honors process, appreciates differences,
makes distinctions, gives significance.
We don't make beautiful and special children
by following the directions on an
instant special kid package.

A Brown Study

Except for a few bright trees, this is a muddy fall. The season falls short. If summer had held its promise, I might not object so much to a brown autumn. But summer had looked and felt like an indistinct haze of grayish yellow. Like a two-year-old who blends all his water colors into one muddy brown, fall seems incapable of making distinctions. The final insult is the persistence of fleas and crickets seeking entrance to the house.

In the miasma of the brown study, some red flags go up. I've been here before and I know very well that unfulfilling seasons and fleas have little to do with the way I'm feeling. Brown has to do with choices. Somehow it is easy for me to plan around other people's needs, after which I convince myself that I'm actually doing what I need to do for myself. This response becomes so pervasive that I have trouble deciding what I deliberately chose in my day – and the brown comes seeping in.

Shel Silverstein has not done much to help this cultural dilemma. He wrote, some years ago, a book called *The Giving Tree.* I imagine he meant to promote a sense of the many ways there are of giving, and, I guess, the

69

value of trees. The book presents one overriding concern for me. If I give my leaves, my branches, my trunk, will I someday end up the stump that is only good to be sat on?

Some trees do give up their lives to become boats, houses, tools, weapons. Other trees just give us the sense of "tree-ness" as they replenish themselves each year and grow colossal branches.

Now, here are the choices: do I want to be all used up, or do I want to hang around to let saplings know what it is to be a grown tree? If I want to teach a child about the value of being a human adult, I cannot demonstrate by my life that being adult is simply being used up. I know with a great deal of certainty that children get a sense of themselves from the adults in their lives. When I take time to pay attention, my child becomes a listener. If I get excited about small surprises, my child senses the joy in life's little gifts and does not need every day to be Christmas. If I value my own talents, my child knows it is fulfilling to be the person I am. If I appreciate ways that we are different, my child values his own and another's talent. If I live each day less as though I were trapped, my child learns that he has choices.

The wind has picked up. The air has turned crisp. Fall was always my time of year. I'm going out to find that glorious golden tree I drove by yesterday. After that, I'm buying a ticket to Friday Night's Philadelphia Orchestra Concert. *Choices.*

In the Middle

There were three of them, by turns upset, explaining, sullen. "She pushed me from behind." "I saw her fall," "I didn't push her." The teacher invited the young children to sit and then began a description of a sociological tool called a continuum line. "In the middle we have a misunderstanding. At this end is a peaceful solution and at the other end is violent confrontation. Now here we are in the middle. What steps will we take next? In which direction will each step pull us, to the violent end of the line or to the peaceful solution? What do we need to do?"

The children went through a series of tries. Just working on the problem appeared to make a visible difference in their faces, body postures, and feelings. They began to relax. It seemed OK to be without an answer so long as they could work toward a goal. World events were discussed: Gandhi's assassination, apartheid, holocaust, terrorism. How did it happen in other places? What was the long-term result? What do you prefer?

In the process the children explored ways of observing, of getting information, of dialoguing. Most of all they learned something about their own way of acting and reacting. The more these children described what they saw and felt and concluded, the more they could arrive at their own style of problem solving, a personal process. What a relief when a child can find inner discipline. The adult, then does not need to make all the decisions, issue an endless spate of orders, and constantly *police* to make sure the child is complying. Furthermore, when I lead a child to work toward discovering alternatives, I learn something new about that child. But the best result in this process is how it says, "You have the ability to understand what you need when I give you the time."

What I do is not nearly so important as how I do it. If an adult wants a child to know how to create a peaceful environment where everyone's rights are important, than that adult better not exercise force to get what he needs. If a parent/teacher wants quiet, he better not shout for it. If an adult wants an honest child, she better not lie to that child. If we don't want our children to grow up to be abusers, we'd better find ways of handling them other than beating them into submission. Our expectations will be thwarted if we have a different set of values for the way we react to the child from the way we expect the child to respond to us. Children learn, not from our words, but from our actions.

We live in the age of instant everything from soup to coffee to housing. At the same time we seem to be saying, "I want something special." Special is individual, is patient, honors process, appreciates differences, makes distinctions, and gives significance. We don't make beautiful and special children by following the directions on an instant special kid package. Special people require brewing and processing time.

Get Out and Up

I climbed the lighthouse the other day. Two days later, still rubbing my aching lower leg muscles, I asked myself why I needed that climb.

Last night I again watched *Dead Poets' Society* and the answer rang in my ears. "When you think you know everything about something, change your perspective," said Robin Williams, the English teacher in a private school for boys, as he urged his students to stand on top of his teaching desk.

Ostensibly I climbed the 199 steps of the light-house to guard a young companion as he reached the wind swept walkway at the top of the tower. Perhaps I actually went to see what I might learn as I changed my point of view.

So what about this lighthouse? Lighthouses, as a group, are obsolete except as landmarks of times past. They have become man-made dinosaurs. I can marvel at the size of this lighthouse, its height, its many steps spiraling incessantly upward, but boats don't need it as a point of reference any longer. Technology has taken away that function. People still climb it and raise funds to keep it in good condition, all the while knowing its use is real only in the memory.

In their dedication to protecting the lighthouse's existence, local citizens also know this. They don't pretend the lighthouse retains its old use. Still, at night, the beacon flashes across the sky, and one remembers how it was when sailors used it to signal the channel fifty feet from shore that meant safe waters. Technology leads the way now. The old lighthouse remains tough, beautiful, still functioning, but virtually useless as a present-day point of reference.

I wonder how I am like this lighthouse. How do I process those once useful, hard, yet beautiful ideas whose time has passed? How do I test and recognize what is obsolete and what is still good for any time and any place?

In a world where the latest computers become obsolete in less than six months, where engineering professors in leading universities can no longer teach many of the skills they once learned because they have become obsolete, what can I tell children that they must learn?

The best place to begin is to understand who I am in the midst of this speedily changing world. I need to meet change without undermining the foundations of my knowledge and my values. If I trust myself to understand what I need, what I feel, what I think, I can make sense out of what I can do with change.

Connections

Children don't wait, can't be hurried,
and will experience with or without our awareness.

... Sometimes we deliberately look to our child
for refreshment.

It Comes First with the Light

Gradually days lengthen and the slant of sun's rays becomes more direct. Then some warm weeks fool the bulbs and bushes into thinking that light and warmth mean spring. Green shoots emerge. Desire to see the first blooms and the first leaves, becomes my preoccupation with even my most casual glance outdoors. Spring will soon be here.

Isn't this always the excitement, the firsts – first signs of spring, first smile, first words, first steps? One anticipates the coming stage and looks for the marks. Sometimes the wind blows cold again and slows down the process. Would-be premature flowers are held back. It's not the time yet. It's too soon. First blooms don't make spring anymore than first words make a conversation. Amazing how we can know this about seasons and walking and talking, but find it hard to believe about other processes: building relationships, learning new skills, even learning to read. We seem to think that first means best and forget to wait for what is lasting.

Being first, doing the most, having the greatest speed, are hallmarks of our place in history. Considering this, the award-winning film, *Driving Miss Daisy* is a throwback. Audiences are brought to a screeching halt to hear a new old-story. It seems incredible that one of the most acclaimed movies is not one that flashes, crashes, glitzes, or screams through outer space, but is rather one that pursues the hidden interspaces between persons.

The time is the 1950's. Daisy Werthur, an elderly southern lady, is having a hard time accepting the limitations of aging. When she backs her car into a ditch, her son insists on providing her with a chauffer, Hoke, a negro, retired from the Werthur factory. Miss Daisy is cantankerous. Hoke waits patiently. They are now destined to journey together

Quietly, *Driving Miss Daisy* shows how attentive waiting leads to *being in tune with*. Miss Daisy, an old teacher, shows Hoke, an old student how to read. Hoke an old student teaches an old teacher how to wait for friendship. Together they learn how to celebrate a life and how to close a life. Seems that this is what all learning is about.

"Did you make any friends at school?" we ask the child, as though building a friendship were microwaveable or done in a day. "What did you learn today?" as though a skill does not need to be built over time and with practice. And with every question we make a statement, send a message, about what we expect and what we value.

Maybe I need to ask, "Who helped you today?" or "Did you have a chance to help someone today?" How else will a child learn about first flowers, full-blown seasons and the seasons to come? If I value another's timing, practice periods, and style, I say that each way of getting there can be valuable and can be different. Finally, I need to take the time to get there.

Our culture has discarded so many rituals, yet rituals are those ceremonies that take us from one place to another. They help us to make the bridges, the transitions. They involve others. Sometimes, as in *Driving Miss Daisy*, the one who helps is a stranger, not even of the same race, religion or social position. And often the one who helps is also helped. The lesson is taught that will take us from the first blooms to the end of the season.

What's New?

It was the first cool day of fall, and I was already thinking about the holidays that would come too soon. Mail order catalogues seduced me into pouring over their contents thinking I'd actually find something new for friends and relatives for the holidays. Nothing seemed new: everything seemed a glitzy rehash of old ideas. The mall, too, presented nothing different. I felt bored and exhausted. "Maybe," I thought, "I've really experienced too many holidays. That's why holidays need children. They're surprise-able. It's the first time around."

I left the mall happy to get out to the fresh air of the parking lot. The noon whistle blew its stale air across the wind, and just then I noticed a young mother and her child. She, probably in response to that twelve o'clock whistle, was pulling her lagging two-year-old down the parking lot. Her head and eyes pointed ahead, drawing her body to form a declining V with the opposite slant of the child at the end of her arm. Just then a gust of wind swooped over him and lifted his fine, straight hair, like the silk fringe of a lamp, all in a wave. As if he needed to hold his hair on, he clamped his free hand to the crown of his head. All real fear of baldness was denied by the music in his high-pitched, "Wheee!" His excitement became mine, but his mother missed that refreshing moment.

Parenting is a tough job, demanding of mind, body and spirit. It's an impossible job if I do not have a way to be refreshed. When I think that all I have time for is doing jobs, I need more than ever to look at the child. Children don't wait, can't be hurried, and will experience with or without our awareness.

Sometimes we deliberately look to our child for refreshment. We ask, "What did you do today? A child won't find it easy to respond if I'm not in the habit of sharing the present with her. If, on the other hand, I am first to offer a story about my day, simply out of the desire to share it with her, my child will more naturally do the same, without an invitation.

Sharing interests connects us to each other. A parent asked me recently how she could get her four-year-old into dancing without having the child feel as though she were being pushed. "I would like her to dance, but I'd like her to do it because she'll like it and have fun doing it." Because one good question so often provokes another, my response was, "How would she know dancing is fun? Do you dance?" Then she offered that, though she was not a dancer, her mother was. "Has your daughter ever watched your mother dance, or watched you responding to the music and rhythm of other dancers?" When our behavior shows what we enjoy, we say it all. We don't need to talk dance, we simply need to share the experience. Almost every man, woman or child, wants to try what they see others enjoying. If your child senses you enjoy sharing your stories with him, he will be motivated to do likewise. If a child watches adults delight in the dance, he will want that same delight.

Value the stories of your life enough to share them with your child. That sharing becomes your invitation to her to join in your world of experience. Your child will reciprocate, and the well can be replenished many times even in the midst of a busy parent's day.

The Old Cane

She was standing in the middle of the long country driveway, hand raised to catch my eye. She was too short to be noticed amongst the tall waving weeds. Her real estate mother had positioned her as a lookout for us, but I had driven past. Undeterred I drove down the same stretch of road again, this time paying closer attention. Mission accomplished! I found the property.

A bumpy driveway brought us the two hundred feet up from the road. As I got out of the car, the feeling of the old place greeted me. I was no child, and I had often felt the *spirit* of an uninhabited place, but this was the first time a long-empty house, and the acres around it, spoke so intensely to me.

The house and the grounds said *life*. Outside between the double windows facing west, a rusty metal cup holder still clung to the wall just above the hose bib. Inside various cabinet drawers yielded holy picture cards, packets of extra long bootlaces, an assortment of keys, and even a man's new white socks. Off each side of the main structure were two low, slope-roofed additions that were clearly the domains of the household seamstress and cook. In the basement between thick cement walls were row upon row of preserved fruits and vegetables, tangible marks of the last man and woman to live here. On a hook behind the porcelain-knobbed door,was a slim, simple wooden cane. I could work here.

Later, as we began to renovate the old farmhouse for my offices and meeting rooms, a procession of former occupants gradually poured through the house.They were curious, but they also wanted to own their piece of the house's history - their claim on what was no longer theirs. They filled in the stories we had already started to imagine, and added some tales we couldn't have dreamed.

The huge mulberry tree that bisected the upper driveway was the spot they had chosen to chain their old watchdog. His chain was fastened to a metal post driven into the ground. Unintentionally they had made a perfect conductor for lightening. They watched him light up with electricity during a summer storm.

There was a white horse named Silver who now lay buried in a patch of ground that my daughter had already claimed for her new vegetable garden. The men told stories about a living room door, long ago plastered over. That open door had cleared a perfect circular path for brothers to chase each other from the dining room to the living room. I wondered how the ten people who filled this house had enough living room within its small spaces much less any play room. Indoor chases seemed physically impossible.

The eldest son remembered helping his father build the outdoor oven and barbeque. Another pointed up the stairway to the second floor. It held two small rooms. The four boys slept in one room and across the hall, the four girls slept in the second. They seemed proud of those hardships. These grown men and women with eyes that shone with delight in retelling stories of their childhood, had come back to see the new old spot and tell the scary, funny and proud anecdotes of their lives in that place.

Stories are often markers that give evidence of a time and place where one learns to conquer fear, to do a job, and to celebrate both. These are the tasks we repeat over and over, almost as though we have to practice until we get them right. The stories we tell and retell, remind us that we've conquered the monster by confronting him and finally by inviting him out of the closet to rest in the bed beside us.

In the process of making sense of the stories of my life, I save some things and discard others. Some I save in memory, some by repetition, and some as metaphors for life. The old cane I found in that house rests on the wing chair I've picked as my writing space. Sometimes metaphors cross from one life to another.

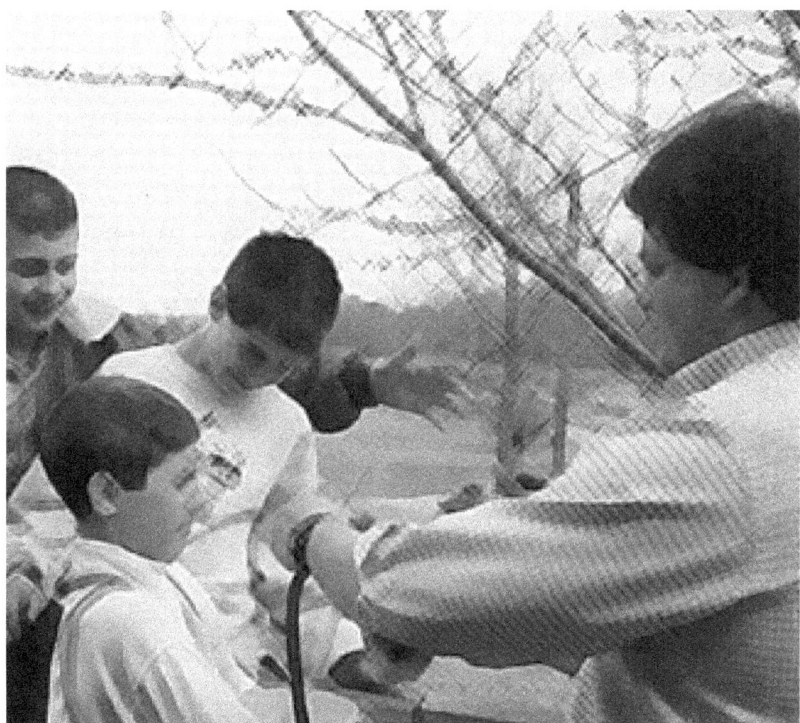

Communicating

A piano's sounding board makes the difference in the quality of the note.

Checking the Sounding Board

As my children grew more verbal, I came to understand clearly that we had shared some of the same events but had experienced them differently. One day, in a light-hearted mood, one of my children and I were sitting around the kitchen table talking about long passed interactions with each other that had always bothered us. When we finished our separate confessions of guilt and bad feelings, we realized our stories didn't match up.

My adult child remembered shared incidents that I couldn't retrieve from my store of memories, and she had totally forgotten the one exchange that still makes me wince at the poor parenting it demonstrated. "But I don't even remember that," was her reply. In turn I reported how painfully inept that memory always made me feel, and how sure I was that she had an internal scar somewhere from that day on. We both laughed over our foolish past.

I am a very different person from the children I teach. My childhood was different. As a child, I can remember roaming my neighborhood with my peers. We knew every field, every change in the ground, each young tree that could be used as a one-person seesaw because it was strong yet supple. In spring and summer, the overgrowths of vines behind our garage, allowed us to weave interlocking tunnels. Away from adult intervention, we learned the power of the group and we designed our first club.

In an attempt to understand the differences in my childhood from those of the children I teach, I ask children today what they do when they're not in school. I wonder how they are experiencing their childhood and their peers. I know malls, sports, lessons, computers and TV are the primary designers of daily *free* time. I know parents are afraid to allow young children to play many places on their own.

I am confused about how children today learn about themselves and their peers when almost all of their activities must be so closely monitored by adults. I am concerned about the values taught by malls and other consumer *parks,* as well as by television. Another thing that concerns me is children having insufficient unstructured time to interact as children with other children.

On one hand, children seem too restricted; yet on the other hand, what can potentially harm them must be monitored closely. Physical safety is a priority, but so is emotional safety which seems a lot less tangibile. Here are some guides worth considering.

> A TV in a child's room needs restrictions.

> *Adult* videos, readily accessible, are like loaded guns. They'll be used, and not always by adults.

> Obscene language that makes fun of body parts

and functions can be sef-esteem reducers. It confuses what is valueable with what is laughable.

What children may have seen or heard can disturb them. Talk to them. Better yet, let them talk to you. You'll soon sense the confusion and be able to understand how to share the experience.

Children have a much shorter history than we have. What they can't understand out of actual experience, they fantasize about to find the missing pieces. Furthermore, they shouldn't have to handle out-of-their-depth issues that an adult can perhaps take in stride. A child doesn't have the same ability to abstract as an adult. Some information they just can't make sense of.

Children need parents as a sounding board, much like good pianos have. A solid sounding board makes the difference in the quality of the note.

Mixed Messages.

Don't Drink and Drive the banner reads as it waves over the heads of the beach crowd. Immediately following is another plane with *Bacardi Bangers* trailing behind it. Somehow I feel as though I am being given mixed messages.

I've become particularly sensitive to this confusion. You know, words that say one thing confused with actions that say another, or two sets of words (or actions) that contradict each other. So I pay attention to the way I shake hands. When I offer my hand I try to make sure I don't stop with my fingertips. The mixed message dilemma is one I'm familiar with. One of my children brought this failing of mine sharply into focus.

Of my four children, this child was particularly shy and quiet. I suppose I didn't think too much about it until her kindergarten teacher said something to me. "I told her to repeat what she said so loud that the boy standing across the street could hear it. Of course she didn't," she reported. Good parent that I believed myself to be, I began to work on getting her to project her voice, to stand up and be heard. Nothing seemed to work.

One day in the midst of one of her first adolescent stands for independence, she and I had a confrontation. She shouted an angry remark at me and headed to the garage to empty a wastebasket. I, in a fury of righteous indignation, followed her, took her by the shoulders and said, "Don't raise your voice to me, young lady." Immediately, I heard myself. For more than eight years I had been trying to get this child confident enough to be heard, and the first time she was, I demanded that she stop. I might have said, "I don't feel right about the way we're talking to each other. What can we do to say what we need without hurting each other?" What I actually said by my words and behavior was that I was glad I couldn't hear her all those years.

The field of education as well as parenting – and politics, and business and advertising and law and medicine – is replete with mixed messages. One of the biggest mixed message bags is the kindergarten *readiness* bag. It can go either way. Sometimes it says hold children back: the older the child, the less chance of failure. Other times it says that development is more important than age, and, by the way, make sure you know your child's gene history to determine whether or not he or she will reach puberty early. Early puberty and being held back can create a problem. Some say test for readiness, while other criticize the unproven validity of those very tests. Added to this are the expectations of the family and the school and the neighbors and the social group about what *should be*.

I get an image of an engine with a start button that lights up when it's time to go. Children are not nearly so simple as engines, and developmental timetables are just tables. A child is a composite of individual variants.

To even the most casual observer, some children give a very clear picture of being too young. Others need ancillary interventions to help their first academic strides. They are not ready to work without constant interaction with an adult. They find it too difficult to move from more vigorous physical play to quiet paper and marker. Physical tasks still involve so much of the whole body that one knows that the child's development is still engaged in gaining control over large muscles. The child's self absorption says he doesn't understand the needs of the group. All of these stand out as signs of a maturity or distractibility level that will interfere with the group setting of a kindergarten. There are other more subtle indications.

I suppose what disturbs me most as an educator is that we put this readiness fixation into a timetable that really denies that children are different, have different paces, have different styles, experience uneven development in different areas. With such an emphasis on testing and scales and ought to's in kindergarten, first, second, and even third grades, we say developmental differences are all right when it comes to walking, talking, toilet training, losing teeth, puberty and aging, but it's not alright when it comes to school. Mixed messages!

Saying all children need to perform according to an educational checklist, does not make it happen anymore than telling a tooth to come in will get the tooth to pop through the gum. And if a child is developmentally different, which may have little to do with intelligence, do we say that his talents do not match what schools want? Instead of wondering, "Is the child ready for school?" I would rather ask, "Is the school ready for the child?"

Hands

I think of that photograph. Standing, looking straight out of the sepia print, are my father, my grandmother, and myself. I was in the middle of the two. I can remember little of what she, who was so old all the years I knew her, spoke. What I can still feel is the touch of her hands, large hands that were heavy on my seven-year-old shoulders. I have her hands, no doubt.

Her fingers were articulate and always in motion. She was the center around which all of us gravitated each Sunday as we made our way to the country house where she lived. From the quiet center that she was, she reached out, not in words, but by touch. The youngest sat beside her rocker in the country kitchen. The other chairs radiated from that spot. She would comment by touch or a single word, "Bella," on what we were wearing as she felt the material and stroked the fabric into place. She would hold my hand in her large one, and rhythmically pat it with the other, all the while smiling. I felt I knew her so well, yet we seldom exchanged more than a word or two, she in her language and me in mine. Her speech was in her smile and her hands.

From that old woman, who spoke no English, could not read or write in any language, we learned the most important lesson of our lives. At fifty-four, she had left everything she knew of her world in Sicily to sail to a foreign country. She had cared for eleven children and their progeny. She was not a woman of means, but she left us a legacy – not money, not fame, but a quality of nurturing that found its way into the lives of her sons and daughters. The unfamiliar ways of this foreign place were taken on by her children. She remained centered on what she knew and valued. Us she accepted without judging. She knew how to love without requiring a payback.

Unconditional love, we call it today: a love never colored by what any of us did right or wrong; a caring that was never withdrawn from us when we did not meet her values or her standards. We all understood what she was, and we sought to match her love. She showed us how to care for ourselves and for each other. We felt our worth from her nurturing.

This woman would be called illiterate today, yet her sons were engineers, surgeons, pharmacists; her daughters, the working force which made their brothers' professions possible in those turn-of-the-twentieth-century days. From her I come to understand the ways we educate ourselves in the real world. Schools have little to do with that. We learn, or not, out of the climate of the spirit. Our schools, if we're lucky, amplify the spirit we already possess.

Separations

Some pretty simple observations tell me it's the end of summer. With only a week left until Labor Day, I alternately say, "Where did the summer go?" and "Thank goodness the heat and humidity will soon be gone!" with "Will I still have time to sit in the sun?"

I vacillate between being glad that my schedule with others will settle down to a more fixed routine, un-influenced by the vacation-pulling call of the sun, and feeling as though I need some quiet, away time, before winter demands take over. Summer will soon end but I don't seem to be finished with it.

I watched my granddaughter this morning as she pulled her stuffed dog along to keep her company while she filled in time waiting for her mother to be finished with some office chores. First she tried to interact with the adults, to keep them actively involved in her play/work. Her mother reminded her that she had explained how it would be this morning, and that if they played to-gether there would be not much use in their coming into the office to do that. "I need to work this morning, and you need to play by yourself for a while." The positive voice said more than that to the child. It said between the lines, " . . . and I know you can do that." The little one gathered some playing cards, some colorful counters, two other stuffed animals and began to organize play that I could hardly understand but which kept her totally absorbed. She invented, dramatized, checked out various rooms bringing them into the scenario she had created.

She lost her mother as a playmate, but she found her own work, the work of trying on characters, moving through space, and feeling good about what she could do for and with herself. Losing her mother's company for that period of time allowed her to find some things out about herself. Like all of us, initially, the child does not always want what she needs.

I've interviewed numbers of parents, students, and teachers these past months. What I'm feeling right now seems to echo what most people experience during transition times. "Now that I have so little time left here, what do I really want to do?" Seems as though a sense of losing something focuses me on what has to be done, as well as what is now important to do without. What I need from a season that will soon be over, a job or a home that will soon be left, or a child who will soon leave home, becomes less cluttered with non-essentials.

A child can believe by the parent's tone of voice and physical stance that she can function without that adult's constant attention. She can also feel that sometimes it is all right for a parent's needs to come first. Transitions, and the separations they entail, require preparation by those who have been there before us. Adults tell the child that others have done this transition, and the child can trust the outcome. The child comes to know that some things need to be let go of, so that other things can happen.

Parents are not the only ones who worry about separation and its trauma. These past weeks give me evidence that changes in schools, changes in jobs and careers, and just the changes that the seasons bring, can cause stress and anxiety, to say nothing of guilt. Guilt is probably the worst and the most debilitating. Knee-jerk feelings of guilt can often get me to ask the least-productive question, while causing the greatest expenditure of energy. "Why did I?" "Why didn't I?" "I should have?" All seem to generate physical distress, and they block the mind's ability to think about ways to arrive at different choices, options that are not so stress inducing.

Our ancestors knew more about human needs during transitions. Rituals involved community and made life cycle passages less lonely. Parents did not have the sole responsibility for seeing a child through the various developmental ages and stages. Ceremoniesmade parent and offspring feel part of a group,

I've decided to keep the part of summer I like best, the walks along the beach that put me in touch with land and sea and air and the free creatures that fill them. I can do that summer or winter and each season brings its different flavor to that familiar place. I'm going to celebrate that old ritual day, Labor Day. Before the weekend is over, I'm going to gather as much as I can find for a saltwater aquarium and terrarium. I want to share some of what I love about summer with those who join me in my winter walks in other places.

To Risk the Dream

*How tenaciously she held on to the dream for each of us
until we could finally hold it for ourselves.*

The Dreamer

The world comes to her now, to her chair, to her bedside. Letters, visitors, mementoes, gifts go into that small room where she lives. She is almost 100 years old, and this room has sheltered her for more than ten of those years. She used to say, "I'd rather wear out than rust out." She has done neither. The body is weakened but the mind and spirit soar. Not even this small room can confine or dull her.

When she writes, her hand is still firm. Her script flows like her thoughts. To this day she has lost neither the vision nor the enthusiasm of the dreamer. She dwells in memories only long enough to recall a specific talent or some timid inclination in others that she can use to inspire them.

I remember the boys in high school as they succumbed completely to her ability to know them better than they knew themselves. We would go to the high school hangout and talk about her over glasses of coke. We paid attention to what she said because she fascinated us with ourselves. "How could she know?" we wondered.

She was the quintessential teacher, invested in drawing along the depth of possibility in each of her students. How tenaciously she held on to the dream for each of us until we finally could hold it for ourselves.

This old teacher's style was a mixture of inspiration, some perspiration on our part, and memorization: *"Music, when soft voices die, vibrates in the memory ..."* She seemed to scatter the contents of her subject over us. I often wonder what followed what. She drifted between her senior classroom and the library she monitored. That last charge kept her at school for hours after everyone else had gone. We always knew we would find her there. And the books and the tasks she put into our hands had a life hidden in them.

Forty years have passed, but I can still smell the newspaper ink from the presses in the next room. Max Leuchter's solid, focused presence met me across the large desk. This impressive, yet soft–eyed editor was patiently interviewing a very green college student sent on an errand by her high school English teacher. She was that teacher. My task was to convince the editor of the paper that I could write. I wrote for him. He corrected my spelling, and then gave me my first by-lined column.

Dreams are personal, individual, and idiosyncratic in every way. They are fashioned out of the most intimate essence of what we are. They make learning and living possible, even up to a hundred years. In fact, everything outside the hoping that the dream brings belongs to someone else. The dream is ours. It's the blueprint of our tomorrows. When parents and other teachers can understand the function of dreams, we will no longer need to worry about turned-off, jaded, or dropout children.

The vision of what can be is informed by a child's instincts about the self that no one else knows. The child's instincts speak and draw along his strongest efforts toward the vision the dream creates. When a child follows her dream, she comes to realize the purpose of her life.

What excitement that kind of learning brings. Productivity and creativity follow naturally from the dream. Finally, there is no one more powerful than the dreamer and her dream can fuel us for a hundred years or more.

What They Long For

Children often talk about what they love, and what they wish would happen. When they describe how they do something they enjoy, I can get a sense from their eyes and the remembering expression on their faces of how that is for them. The same expression holds when they talk about where they would like to be, and with whom they need to join to do special things. Given the chance, and a listening facilitator, children can be amazingly adept in discerning what is good for their own best interests.

I wonder about what it will take for them to reach what they long for. I think about what they will need to practice to be able to hold on to their dream. One thing I know. As they grow older, they will find it harder and harder to hear their once true young voices over the noise of others' *should's* and *ought's*. Gradually they might scarcely be able to tell their own sound from all the others who advise them *for their own good.*

When children are not taught to value their ideas, dreams, fears and feelings, they learn to be tone deaf to the sound of their own voices. Dreamers, creators, leaders, die when they can no longer hear themselves.

When we interfere with a child's dream, I wonder what we fear most. Is it that our children will not be like us, or that they might be too like us? All the cajoling, manipulating, or even threatening maneuvers we practice carry overtones of lack of trust. They say, "We can't permit you to act on your own behalf. We already know how terribly misleading self direction can be." If we were not believed as children, we grew up to distrust our best instincts, and to look skeptically at those instinctive behaviors of our children. In the long run, what we fail to trust, we debilitate. Not much can happen when one feels weak.

What we do need are some skills in how to listen to our children without preconceptions, expectations, assumptions, panic, or a trail of our own personal garbage out of our own bad feelings. Look at the fresh heart; respond to the different voice. What wonderful things can be discovered about this new person who happens to be your child? Then, allow the difference to happen. Dare to go to the edge. That's where the dream is.

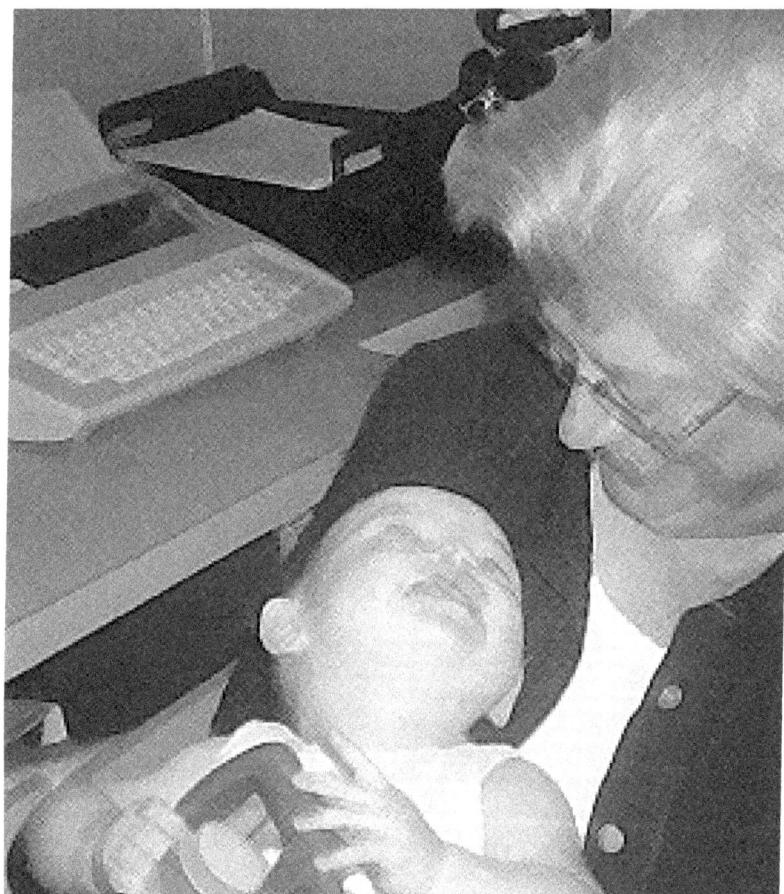

Eyes and Ears

You are learning what it is to be alive.
I am relearning the simple tasks I often assume
are always in place
just because they were once learned.

The Past for the First Time

Slight and hunched, like someone permanent-
ly bent over to hear, she finally raised her head as she
reached the podium. I was amazed that the voice from
so frail an image could vibrate such deliberateness. She
had taken the stairs a slow, solid foot at a time. Yet except
for the slight defining pause as each foot contacted the
wooden floor, she crossed the stage with a step so shallow
she might have been gliding.

I expected the bare and unadorned stage to rob
any warmth from her voice. The cool, castle-like theater might deny us any intimacy with her. But her face
matched the vibrant pulse of her southern voice to reach
us like a strong, out-stretched hand. I don't remember a
word she said, but I can still feel the quiet center of her
voice like some magnificent listening ear. Though she
spoke and read, I heard, not her words, but her listening. I fantasized that even as we sat before her, she could
hear the *eye* of each of our stories. This was a fantasy fed
as much by the power her writings delivered as by her
physical presence.

The Eye of the Story was a gift to me from a friend intent on responding to my first interest in her writing. I longed to share the book, once read, with everyone working in human development. My message would be: "Look. This is what writing can do for the writer!"

"I don't know her," a colleague said as I sent a copy of her book to him. "What's her name again?" "Eudora Welty," I answered, only a little surprised that this Pulitzer Prize winner could still remain unknown.

This night she returned to Bryn Mawr, where she had taught. A slim gray book, *One Writer's Beginnings*, had just been published. It was her latest following a list of stories, novels and essays. Eudora Welty had begun her autobiography. The chapters, only three, were titled simply "Listening," "Learning to See," and "Finding a Voice." Not so simply, her masterful hand intermingled those three skills, and allowed each sense to be used interdependently. Listening is practiced when one finds a voice. At the same time the fictional eye feels. Whichever sense, it was nurtured by her insatiable need to know, and everything was used.

One Writer's Beginnings, unsurprisingly, reveals more of the life traveler who happens to be a writer, than it does of the private life of Eudora Welty the writer. The private person I am, delights that a writer can reveal so much while telling so little. Nor is her autobiography written to place events chronologically. Welty, looking back, hears the story in the memory. *Revelation* provides the *continuous thread*, and what Eudora Welty had done so often for me as a reader of her fiction, she now provides for me as I read her life. I know it is this *thread of revela-*

tion that draws me to her autobiography rather than the thousands of others I might pick to read.

She was in her seventies, but the old memories she used were not just the endless repetitions of old age. She wrote as though she had discovered her past for the first time, like unread pages of an old journal. She saw and heard the *whole family life* removed from the time frame in which she first experienced it. She became her mother and her father, and then created herself anew. Her desire was to discover that long ago past to recreate herself. She paired what is seldom joined, youth and wisdom.

I comb through her words, the way of her experience, the poetically terse, yet rhythmic wisdom that refreshingly tells me more about *how* than *how to*. In an attempt to revision my past, I practice writing in another way hoping to find what is new in an old life.

My Voice or an Echo

As soon as spring declares itself with whatever weak little blossom or green sprig, thoughts of the beach begin to intrude. A sunny day is hoped for, but rain is no deterrent. The call to feel again that stretch of sky and water will not be put off by the weather. Spring, with all its announcements of beginning again, seems to prompt this need in me for a place that un-wrinkles the mind, tunes in to the body, and charges the spirit.

These weeks I watch the children as they too become sensitized to the new light and colors of this season. They alternate between high, energetic excitement and lethargic, listless, almost turned-off behavior. "Spring fever," we all say, as though some sickness were involved. Yet, I suspect this fever, like other fevers, is the body's response to a need to renew itself. Children don't fight these natural responses as much as we tend to as adults. And there is something to relearn by paying attention to how the child asks for what he needs.

A very young child who had been sleeping well nights now cries at bedtime, wakes often during the night, needs to be with a significant adult. As soon as

116

the adult lends a calm, soothing, almost healing presence, that child's misery subsides, sleep returns, and the body can trust itself to rest again. One might be tempted to say the child is demanding, manipulating, and on the way to being spoiled; but what is the message we send that child when we leave it to cry its way through some agonizing nights? Often it is this: "Your experience is not valuable because I know better. There's nothing to be afraid of. I know what you really need."

A concerned parent wants to be sure he is doing the right thing for his child, however, interactions between the two generations can take on the flavor of test time. In this mode the parent holds all the right answers and the child has to match up. Subsequent exchanges become loaded with anxiety on the part of the parent, who worries that the child will not pass the test. Of course, under pressure, the child falls short of the parent's expectations. The result - the child gets to measure herself, not by learning to trust her experience of herself, but by taking her temperature in relation to the adult tester. Such a child fails to acquire her own voice and instead becomes an echo. That parent does an effective job of teaching the child to deny her own experience.

If I teach myself (and in so doing teach my child) to pay attention to what the mind, the feelings, the body are saying, I have taught the first lesson in leadership. The essence of leadership, after all, is to be able to use oneself as the point of reference in any situation. A leader learns how to value his own experience and honor his own history. He does not simply use the external histories of those around him. A true leader knows his own voice from an echo, and can use that voice to make a difference for himself and for his world.

Spring is renewal time, time to get to the heart of where all growth happens, time to value the deepest possibilities that emerge when I honor what is distinctly yours and uniquely mine. Spring for me is the time when I need the vastness of that seaside space to clear some significant inner space that allows me to be in tune with myself and those I meet.

What's the Payoff

Most parents work pretty hard at their parenting jobs. They ask, "What did you do today at school?" They cart children around for all kinds of lessons and team sports. They appear at matches to cheer their offspring on. They give birthday parties, sleepovers, cookouts, and pool parties. They make various trips to known places of educational and recreational importance. They make sure they have enough friends available. They supply their darlings with all the popular games and toys, clothes, and the latest computers. Some of them work two jobs to do it. And after all this, there's college.

Why do parents choose to be parents? What's the pay off? Certainly parenting is more than chauffeuring bodies and paying bills. It should be more than worrying a child through a given number of years. What's in it for the parent? What are the joys of parenting?

In some quiet moment, I suppose a vision exists of what a parent is and what a child is and what it is for the two to be together. Part of that picture might include companionship, shared mutual interests, and a parent's ability to pass along some important stuff about living. We've learned something and we want to share that with our child.

Maybe one might even dare to imagine another scenario where a child can eventually progress through the my-mom-and-dad-can-do-everything stage, past the they-don't-understand-me syndrome to the awareness that I-have-real-people-for-parents stage, and I like them. This last level of parent-child relationships is perhaps the most frightening, and the most rewarding. One leaves off playing the guru parent role and instead can exchange ideas, opinions, feelings and needs. For this to happen, someplace along the way a parent had to extend the invitation to walk side-by-side.

It is hard to give up the adoring look a child directs to the imagined all-powerful parent. It is distressing to feel inept while one watches a child struggle alone because that child needs to practice what it is to be a separate person. It is most difficult as a parent to admit to feeling less than OK, in charge, and in control in order for the child to understand that strength is borne out of enduring pain. But sometimes, less is more, and in giving up the power in the parent role, one allows the offspring to be empowered.

Being a parent can sometimes get in the way of being a person. But of all the skills, a child needs most to experience what a mutual relationship can be. Parents can model that, not only with each other, but also with their offspring.

I hear you listening.
My rush of words
Become too much and
Not enough.

Echoing the physician,
I take the hero stance,
Cut through the sensitive
To search the vital.

But heroes take the step
Beyond and I would walk
Where I can see your
Face.

I open my hand,
Let go,
Wait for
The dark, inexpressible, beyond reach, unhurried
Deep
Where we unfold
Together.

You're Just Two Now

You can't read or write. There is little in your life, short as it is, that you can remember. Your aunt has pictures of you in last fall's leaves. You look at the pictures, but today's leaves are all you can see. You seem all present tense. I guess that's why I'm so surprised at what you figure out and put together and then verbalize.

Saturday, when you sat next to me and showed me your lollipop you asked, "What flavor is it?" I marveled at how your developing brain can move from color to flavor. I pressed you further when I answered, "Lemon. Does it taste like a lemon?" I wanted you to make the connection, to know the real from the facsimile.

Friday afternoon you sat beside me for two hours with intermittent jaunts to get juice and cookies, eat a piece of fruit, or check on your mother who was working in the next room. You were working too. You imitated our pen and paper tasks by filling page after page of small memo papers with your ballpoint markings. When the rhythm changed and you sensed work was soon to be over, you questioned what was going to happen to you next by asking, "What I'm going now?" I'm fascinated by the way you ask the question with so much implied, *give me the words so I can have some control.*

Control is what you've been about since birth. When you were three months old, I was intrigued as you struggled to understand your own hands. Moving the large muscles of your arms, you first had so little control that you startled yourself awake with your own spastic movements. But finally, after much practice, you could hold your own hand.

I don't suppose these little anecdotes will mean much to you for lots of years; maybe not until your hair turns gray like mine. But these small stories do teach me when I stop long enough to pay attention. They remind me that it's a life-long necessity to ask what I'm tasting or smelling or feeling, or even doing. That's the way I keep sharpening the skills I need to tell the real from the make-believe.

When technology will have done another half century of somersaults, when you will be able to read and write books, and work on technology that we haven't yet dreamed about, you may get some old lessons from your granddaughter. Her focus will become your own. That never seems to change.

For now, you are an avid people-watcher, and you catch every nuance in an interaction. You can anticipate that something is going to change even before it happens. When I watch you, I see how I have lost that skill, even after years of practice. You are learning what it is to be alive: I am re-learning the simple skills I often assume are always in place just because they were once learned. It takes lots of practice to hold your own hand.

123

Inside the House

The day is cool, damp, cloudy, one of a series a similar days. As if on cue, physical ailments seem to multiply in direct proportion to the number of leaves on the ground. The cough that says there are irritants in the air has become more insistent and less intermittent. The body seems to rebel both physically and emotionally to the onset of the season of long nights. Maybe this is about change of focus.

One begins to crave hot cocoa, a warm fire, engrossing books, good theater, inspiring music and powerful art. The body has been nurtured by nature in summer; now it seems to need to create an inner world of warmth to prepare for winter. One learns again what it is to know the inside of the house.

Yesterday I went to a distant bookstore, one I had not visited for months. I was not there long before I noticed that the so-called *self help* shelves had tripled since I had been there last: *How to Overcome Stress, Games People Play, and Body Language.* Many titles were familiar, recycled from past years. Many authors were familiar too, with new titles above their names. The only specialized section that seemed larger was the children's shelves.

Vendors cater to the demand. One might question: how does it come to be that people are buying more books to understand themselves? Do books on children perform the same service, and if so, how?

I smile as I remember a satiric line my wise nephew used to describe the spawning of how-to-books: "Get control of your life. Let us show you how." Do we often give over the controls to the experts in our effort to get control of ourselves? Such a paradox! Someplace in our education many of us have not developed skills to understand ourselves sufficiently, and then, to understand others. Now, book titles tell us what we may already know. It is important to recognize the landscape of your inner house, your self.

How's are important. *How to's* seldom work for long. The trouble is that someone else's *how to* formula is the same for all of us no matter what our individual differences. Wouldn't we be more successful if we copied a good farmer? We could use the same powers of observation for the inner landscape that a good farmer uses to understand the outer landscape.

An experienced farmer takes the time to look at everything, to use his tactile senses, his olfactory senses. When he tells us something we haven't been aware of, we tend to glibly say, "It's farmer's sixth sense." We might be more accurate if we said he was using all five senses. What a farmer intuits is largely informed quite simply by what he has understood through his senses. He then uses

125

his experiences to understand that information. I can be an expert on observing the inner self by using this same process. I will not be such a mystery to myself if I can learn to give each day enough importance to remember in some kind of record what the day has been.

Children can very quickly take to keeping a journal. Research has shown that children enjoy writing about how they are each day. Not surprisingly, children also elevate their self-esteem when they keep journals. Certainly journal keeping is a way of taking hold of the events in a life. Children have few opportunities to exert control in appropriate and productive ways, and self-esteem is about feeling in control. Writing about the events of one's life gives significance to the story and control to the author.

Writers, composers, and artists know about observing the shape of a life. When they share their personal explorations through their compositions, they give us insight into the universal feelings we have in common.

That Moment

Some incidents remain so vividly in my mind that I can remember not only the words, but also the exact place, the time, the lighting, and the smells. Every sense is stamped with the importance of the event. The presenting event is not a prolonged interaction, just a few words, but those words say, *Something significant just happened.*

"Now that there is someone who listens, what do you want to say?" he asked. I was speechless. I had watched this mentor's listening before. I understood that I was more careful when I spoke to him because of the difference in the way he paid attention.

Not many persons know how to listen to more than words. He understood that persons tend to tell stories obliquely, on the bias, almost as though they would have to be read from the corner of the eye. He was prepared to see our stories in the images we used, in what we said *between the lines,* in our silences, in what we wore, and in what we chose to avoid.

As a result, his hearing required me to pay more attention to myself. I tested my words against his listening. "How will he hear this?" I would wonder. I began to place a different kind of significance on what I said and how I said it. " What kind of stories did I tell? "Which stories did I place side-by-side? " What did they mean in relation to what I was working on?" "What did I choose to look at as I was talking?" So when he asked his simple question, I was riveted with the awesome power of the listener who, in turn, empowers the speaker. There is nothing like having been listened to so closely. I learned how to pay attention.

Parents and teachers talk about children's listening skills and their ability to verbalize. If I kept a record during my years in education, I would undoubtedly find that the two phrases used most frequently by adults in interactions with children are, "Listen!" and "Be quiet!" Their complaints about children most often seem to be: "She's always talking," and "He won't talk. I don't know what he does all day."

I'm convinced that children, in fact, most of us, want to be listened to. We crave that. In turn, we all seem starved for stories about others. The ability to listen to a story needs little teaching. On the evolutionary scale, listening to stories is one of the skills humans first develop. So-called *primitive* tribes always did, and still do, use the story to deliver the message.

Somewhere along the way, we seem to have misplaced that storytelling skill. We have replaced it with what some might consider more direct, up front, efficient ways of communicating. With children, and even with other adults, we tend to teach rather than make the contact. If we can't tell them something or make the point, we think we have failed to communicate. The opposite may be so. In reality, the power of suggestion, especially the suggestion in the story, is the greatest power of all.

A child says so much in his meandering speeches. He gives a sense of the movement in his thoughts. If I can listen for the story, not only in what he says, but how he says it, that child's whole world can open up for him and for me. I give significance to what the child says. This becomes the moment that is never forgotten, the moment that will reach down through generations. It is the moment the child will remember as the time when he was listened to so well that he heard himself.

Sharing History

*. . . not so much set memories of special events,
but rather a weaving of daily patterns.
Children may not always remember what we did,
but they will always feel how we lived together.*

I Had Trouble Holding on to Summer

Either summer didn't seem to arrive fully or, before I could find it, it was gone. I don't feel as though I experienced summer at all – or hardly at all. Now the airways are filled with Christmas music.

Some young hopeful on the staff of the "Farmer's Almanac," trying to vie with FDR for Thanksgiving fame, took a popular poll about moving Thanksgiving Day to the 15th of October. This would coincide with the harvest nature of the holiday and give people much more time between Thanksgiving and Christmas – ". . . time for merchants and shoppers," he elaborated. Also, he predicted, travel would be easier in warmer weather. An historian from the Smithsonian Institute argued that a tradition becomes significant by being repeated at the same time.

Seems as though Thanksgiving isn't that old, as traditions go. FDR mandated that it be celebrated on the third Thursday of November rather than the fourth Thursday. The tradition really only has a few decades of history – not even one lifetime's worth.

I suppose it really is important to me that some things in my cultural environment remain the same –

133

same time, same place. Having my calendar gears shifted to accommodate a new date to celebrate seems unnecessary in light of all the other changes I'm asked to make. But what I'd really like to influence is the change in the way of celebrating the day – not the turkey and stuffing, not the varieties of potatoes, not the pumpkin pie, just the franticness. A way of slowing down the rush is what I'm looking for.

Thanks seems one of those quick, glib things we learned to use at appropriate times as kids. So often now it loses any real feeling and with that any real meaning. Ask a classroom of children what they're thankful for and the answer sounds like a Hallmark card. This year I think I'll ask, "What is it that you appreciate so much that you wouldn't want to live without it?"

If I can slow down the verbal rush, I may be able to hold onto Thanksgiving even longer than I held onto summer. If I go more slowly, I can go more deeply.

The Heater Is on in Earnest Now

If you stay in this town, this time of year means four more months of squaring your jaw to meet the damp cold. In an attempt to confront this darkest season of the year, our long, long, long ago ancestors made holidays. Though spiritual in origin, they were wisely placed to answer human needs at the changing of the seasons. At this time of the year we need light and bright. Even though we now have the technology to illuminate outer space, we tend to carry this bravado against darkness to a frenzy in our attempt to celebrate with lightness, brightness and gifts. Just one month of this thrashing about leaves us needing the next three months to recuperate from the colds and flu fueled by the stress of celebrating.

Whether you've been a classroom teacher or a parent-teacher, you can tell the season by the pitch and intensity of the sound and movement of our children. From pre-Halloween through New Year's Day, excitement is high, relieved only by the sheer brilliance of the adults who can still hold a child's attention in the midst of the excitement of their anticipation. Even if you're not anywhere near a young child and don't have a desire to remember why we have holidays, a walk through a mall will remind you. The merchants will tell you what to remember. Magazines call it "building tradition."

For a long time I've been trying to separate tradition from meaningless rituals that really belong to someone else. Spontaneous merrymaking that comes from a special personal need to rejoice, a *one shot* celebration, is easy to understand. When it comes to traditions, however, nothing is as easy as it used to be.

Tradition building must have been on my mind the other day when I asked some young children, "What is the one thing you would never, ever want to be without?" The responses were varied, as might be expected: family, toys, even books. The one choice that kept popping up like a refrain amongst these was *my room.* I was at first quietly put off by that answer, but the positive and repeated words gradually began to feel like a metaphor for something they could not express. I began to connect with what the children were saying.

My environment tells me about myself and helps to give a sense of place in this fast-moving, ever-changing world. That is also what tradition tries to do. It seeks to give a sense of where I belong. *My room,* in that same way, gives me a feeling of in-place-ness. I can move anywhere and still recognize *my room,* for it speaks to the core of what I am in relation to the significant others in my world. *My room* is where I can go to listen to the sounds of my life. When I go home for the holidays I reconnect with the space that gave me my sense of place. I touch base to feel my beginnings now that so much else has changed.

I need to get with the season. The mall that underlines what I must wear or own is not the place I need. I long for my personal sense of continuity, the tradition I build for myself. Perhaps if I go outside into the darkness of a December night to see the stars, at their most spectacular, I can find my place. I remember that I can enjoy the stars without wearing or owning them. I can reconnect with the essentials that do not change. When the first snowfall begins, I'll marvel at each flake, knowing that I cannot own these either. I can greet the season in the same way that I can revisit my room.

Sometimes my room means a bookcase that holds things that are precious only to me. Sometimes my room is a suitcase of memories that I take with me wherever I go. Even if that room now only exists in my memory, it continues to describe my sense of place. That will not change. I understand what the children have been saying. A child needs to be able to feel a sense of place, when so many things around him change.

By Mid Afternoon the Day Seems to Close

Shadows are long and distorted against the build-ings. The quality of the light weakens and reminds me that nightfall approaches quickly at this season. Fall seems to rush, not only the daylight but also the weeks and months, and I get to feel as a fallen leaf hurrying along before the wind.

There are many markers in this season, calendar holidays, even a birthday. In the moving car speed always seems to be exaggerated by the number of trees or poles on the side of the road. Perhaps seasons are the same. The more special dates, the quicker time passes.

From time to time I need just to sit and let memo-ries remind me of where I've been down the years. What I remember at times like this is not the catastrophic events, or the special occasions, or even life's most embarrassing moments. I remember bouncing a ball against the side of a house while I wait to be called in to dinner. Not any of the many ball-playing times of my childhood, but that specific summer afternoon. It's as though the full sce-nario, complete with sounds and smells and feelings, gets played out again.

There are many of these vignettes that seem to float on the top of my memory pool. When counted against a full lifetime of happenings, I wonder what is it about these seemingly insignificant events that cause them to remain in the foreground of my mind?

I often look at the children I am with during my working days and wonder, "What will they remember of all that we plan and do?"   I feel sure that there is no way of predicting what each child will keep; no way of knowing which memories of that totally formed up kind will be significant to each one later.

If I look again at the scenes I select out of my storehouse of personal pictures, I realize they are the ones in which I had some private dialogue with myself. I seem to be apart, watching and at the same time doing, whether it's the game against the side of the house, or practicing turning and skating backwards on an early evening; or noticing the light break into the color spectrum as it passes through the beveled glass of our front door. What emerges out of the memory of each of these ordinary, solitary pastimes is the texture of my life – how it feels to be me. I have an idea this happens for all of us in pretty much the same way.

What we make, then, for our children is not so much set memories of special events, but rather a weaving of daily patterns. Children may not always remember *what* we did, but they will always feel *how* we lived together.

PIECES

THAT

MAKE

THE WHOLE

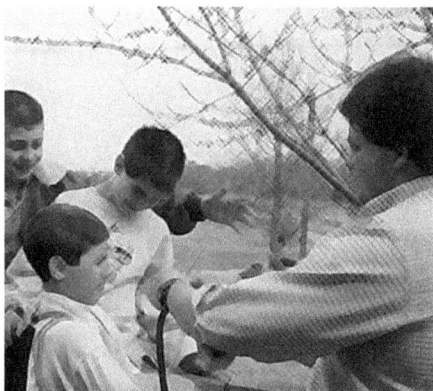

Ann Giacalone D'Ippolito is at the end of her seventh decade. More than five of those decades she has spent in education, teaching all ages from preschool through masters candidates. For twenty-one years she also worked in school staff development in the private school she built with her four children in Vineland, New Jersey. During two of those years, she wrote a series of columns for the town's daily newspaper titled *Learning for Life*. This book is a compilation of those revised articles.

Now retired from the academic world, D'Ippolito lives by the sea at the southernmost tip of New Jersey. She spends her time writing a memoir, working with other area writers, and growing *New Learning Applications, Inc.*, creators of educational products for very young children. Coggols is accessible at www.coggols.com.

www.ingramcontent.com/pod-product-compliance
Lightning Source LLC
Chambersburg PA
CBHW031208270326
41931CB00006B/460